Echoes Of Another

A Novel of the Near Future

Chandra K. Clarke

Copyright © 2020 Chandra K. Clarke.
www.ChandraKClarke.com

Cover design by simansondesign.com
Book Design by www.polgarusstudio.com
ISBN 978-0-9730395-8-0 (paperback)
ISBN 978-0-9730395-9-7 (ebook)

Published by Fractal Moose Press,
an imprint of Tiger Maple Publishing
www.TigerMaplePublishing.com

PART I

RAY

Ray stepped out of the shadows of his apartment block and shivered in the cold.

The sky was beryl blue and clear. The sun was just high enough over the horizon to send brilliant beams of light skittering across the snow, making it sparkle like diamonds. Drifts and rooftop snows had begun to evaporate, wraith mists gently rising into the air.

He tugged his coat collar a little higher. If he had to guess, he figured it was about twenty degrees below freezing. Not so bad, especially for the first week of January, but he knew the buildings lining the downtown could sometimes funnel the cold wind until it howled.

"Summon," he said, enjoying the way his words formed plumes and floated away. The snow squeaked under his feet as he shifted the weight of his short, stocky frame from one to the other nervously. Several anxious moments passed before a single-seater pod emblazoned with the Toronto Transit Commission logo glided around the corner and stopped in front of him. He calmed down a bit, grateful the transponder he'd pickpocketed actually worked. The door opened, and he climbed into the small patch of warmth. There were no controls or amenities, just a thinly cushioned seat showing its age, a restraint that automatically clamped across his lap as soon as he sat down, and a hard, bioplastic dashboard with nothing on it other than an embedded screen. A small heater under the chair blasted hot air at his shoes. There was room enough for him and not much else. His sun visor fogged over from the abrupt change in temperature when the door closed, but he ignored the sudden blurriness, knowing it would clear on its own shortly.

"Destination?" the pod asked, startling him. He'd been expecting to have to tap in a station.

"Uh, Edward. Go by way of Sheppard and Queen's Park Flows please."

The pod nav did a quick calculation. "Estimated arrival time, 7.3 minutes. Entertainment options?"

"Nothing, thanks," he replied. He wouldn't have known what to ask for, anyway. Besides, he wanted to look at the beautiful old university buildings as they zoomed past. Perhaps he'd catch a glimpse of that determined-looking woman with the light-brown hair he had seen there. It would be a good omen, and he needed one. He had a lot riding on getting everything right this morning. This job would change everything.

The pod accelerated, smooth and silent. There were hardly any traffic in his seedy neighbourhood, but as they approached the city core where the buildings were bigger and newer, the trickle of pods became a stream, and then a torrent. He tensed as his pod hurtled towards the major flow that would take him the rest of the way downtown. But the rushing vehicles adjusted, parting to reveal a pod-sized space into which they merged effortlessly. He looked at the dashboard and discovered that the screen showed him his position in real time. He marvelled at the thousands of pods pulsing like white blood cells through the arteries of the city.

Ray tried to relax into his seat, but he was too jittery, and the chair was rather hard and uncomfortable. The pod slowed a little as it neared his destination. Already, he could see people hurrying along the footpaths, their shoulders hunched against the cold. He wondered how many of them were tourists; most city natives knew to use the downtown's underground paths in weather like this.

A familiar face flashed by the window. Startled, he spun around in his seat to look out the back.

"Stop!"

"Emergency stop," the pod replied.

Unfazed by the sudden change of plan, the pod decelerated and pulled carefully onto a footpath, its warning lights flashing brightly. Pedestrians walked around it but otherwise ignored both it and him. Ray wondered how often pods must have to make unplanned stops for it to be so unremarkable. He tumbled out of the pod and jogged back up along Bay Flow, rounding the corner in time to see the man he was looking for duck into an alley off Elm.

"Hey, Mick!" Ray shouted, running faster.

Mick was a tall man now; much different from the gangly, skinny teenager Ray remembered. Mick's long legs and lithe frame meant he took rangy strides and had a fast pace. He was already halfway down the alley when Ray reached the entrance.

"Mick!" Ray shouted again.

Mick stopped and turned, frowning a little. His face broke into a wide, easy grin when he recognised Ray. He waited for him to catch up.

"Hey man," Mick said, pushing a wayward lock of brown hair out of his eyes. He wore a couple of day's growth of beard. "Look at you, all grown up. You're a long way out of J-District, aren't you?"

"I could say the same about you," Ray replied, panting a little after his unexpected sprint. "What happened? You kinda disappeared on me."

"Yeah, I—"

A drone dropped out of nowhere, stopping between them, hovering almost silently. It was big and black and unmarked, and it steamed like a dragon in the frigid air.

It flicked a scanner beam, long and red, a tongue, up the length of Mick, tasting him. Mick's eyes widened in fear.

And then the drone exploded.

KEL

On the other side of Toronto, Kel Rafferty was bracing herself as she unlocked the habitat door.

It wasn't the smell she needed to prepare for. She liked the odours of the hab: the strong scent of wet earth, the powerful stench of the stagnant pond, and the musty, decomposing leaves that permeated air that was so thick with humidity you could almost swim in it. It certainly wasn't like the city outside, and it made guests wrinkle their noses in distaste.

No, Kel was bracing for the macaques, as they would often swarm her the minute she came in, climbing her tall frame, pulling on her long, ash-brown hair, trying to pull it out of its braid or picking at her bag to find the pieces of fruit they knew she carried in there. They were usually gentle with her and took their squabbles to the trees if a fight broke out over a choice bit of banana. Still, it could be nerve-wracking. They were wild animals, after all.

She shrugged off her parka, tucked her hat and mitts inside its sleeve, and draped the coat over her arm. She then let the biometric scanner read the digital tattoo on her wrist, heard the lock release, and watched the door open for her. Kel waited patiently while the hab systems scanned her for pathogens that might harm the enclosed system. Then she unlocked the interior animal security door, the set up always making her think of an airlock on a space station.

When she stepped inside the hab, it felt as though it had swallowed her whole. The heat and moisture enveloped her; her nostrils flared, and she took slow, deep breaths to combat the feeling of suffocation. She checked her wristband and saw it was quite hot, nearly 45°C with the humidex factored in.

As she started down the trail that cut across the narrow end of the hab and led to her office, she smiled and recognised how lucky she was. For as long as she could remember, all she had ever wanted to do was neurological science, and here she was, working at arguably the largest facility in the world for studying animal models of neurodegenerative diseases.

The hab was a fully realised indoor forest, one of the city's best-kept secrets, spanning several hectares in the green zone tucked in between Markham, Scarborough, and Pickering. It was a joint project of the University of Toronto and the Toronto Zoological Research Centre, the latter still known to locals as the Toronto Zoo. Brilliant yellow leaf warblers flitted through the canopy, trading insults and jibes with the partridges that scuttled along the undergrowth. She could hear the gibbons hooting at each other in the distance, and she knew if she went off the trail, she'd see the leaves and plants ripple as the dull brown, striped skinks scattered to seek refuge in the water. Although not as diverse as the Hainan ecosystem it mimicked, the hab contained a wide array of birds, reptiles, small mammals, and dozens of species of insects.

All to support her macaques. Well, not just for her, she supposed. The stricter animal rights laws of the early 2020s had mandated proper living conditions for non-human primate studies, so the administration had shrewdly attracted several other researchers and specialists to the facilities to extract as much value as possible from the elaborate setup. As expensive as it was, the scientist in Kel approved: happier animals made for better and far more accurate data. She also found the other research being done there fascinating. Among other things, they were learning so much more about how mycorrhizal fungi helped trees thrive and communicate.

She reached her office door, unlocked it, and walked in. The cooler, drier air was a welcome relief, as always. The office was spacious, with a wall given over to a large one-way observation window, so she and her colleagues could look into the hab. Another wall was covered with screens connected to more than a hundred tiny cameras scattered throughout the habitat, which they used to monitor the forest and its inhabitants. A door in the third wall led to a laboratory, a pharmacy, and a fabbing area. There were six desks in the office, arranged in an open plan. Even with being delayed this morning — many of the pod flows had been rerouted because of an incident downtown Kel had only vaguely noted — she was the first one in, as usual.

She put her bag on her desk next to a display that shuffled through several images of her favourite grandmother. "Wake screens," she said to the wall, tossing her coat on her chair.

The monitors blinked to life. Her blue eyes flicked to the upper right, where the cameras were trained on the macaques' preferred gathering spot — a small space between the plum yew trees that was good for rolling and tackling and chasing. She furrowed her eyebrows in concern. The animals were there, as expected, but they were agitated, bouncing around from place to place more than usual, hugging each other, baring their teeth, and widening their eyes in alarm. They often fought with each other to establish their hierarchy, but this seemed… different.

"Audio, camera six," Kel said.

The room filled with shrill, anxious, chittering barks.

"Audio off." Kel tried to do a quick head count. They weren't all there, but it was hard to tell how many were missing. And none of them were engaging in grooming behaviours, which meant nobody was ready to calm down yet. What had set them off so? "Status report, most recent log entries."

"Significant event: implant for Max ceased transmitting thirty minutes ago, and he is assumed dead. Implant for Dalton transmitting, with decreasing intensity, from 2.5 metres outside camera sixteen's range."

Kel swore.

SETH

In a modest group of semi-detached houses on Harvie Flow, just beyond Toronto's Corso Italia district, the clock flicked to 8:00 a.m. The window in Seth's bedroom slowly changed from its blackout tint to a beautifully translucent, algorithmically generated frost pattern. He groaned and sat up, blinking in the sunlit patch on his bed. His room was scrupulously clean and very well soundproofed; the only thing he could hear was the filter sucking dust out of the air.

Monday again. Mild anxiety gripped him, squeezing his stomach. Between the surprise visits from his cousin's family and the usual chores and errands that always seemed to take twice as long as they should, it seemed like the weekend had just vanished. He sighed and swung his legs over the edge of the bed, the muscles of his lean, fit body flexing easily, without stiffness.

Seth tapped his wristband, and a virtual screen appeared in the air. He scrolled through its long list of numbers with a flicking motion. Body temperature: normal. Total sleep: seven and a half hours. Fasting blood sugar: normal. He scrolled some more and then hesitated.

"Tasha, why are my selenium levels trending low?"

"Data indicates your selenium levels are within normal

parameters," his digital personal assistant replied.

"Yes, but they're on the low end for the third day in a row," Seth said.

His DPA paused before answering. "Analysis of your diet for the past seven days suggests selenium intake has been optimal, as with all of your other micronutrients, as per your instructions. It is possible your bio-analysis implant needs recalibrating, but records show you had the implant recalibrated last week."

"All right, all right," Seth said, making a waving gesture to force the display away. He walked into his bathroom and peered at his face. First, he checked the area around his big, russet eyes for wrinkles and lines in his dark skin before running a hand over his dense, wiry brown hair, inspecting it for grey. Then he gave his gums and teeth a thorough examination before cleaning them carefully. He went to the toilet, but resisted the urge to check his wristband again for the urinalysis numbers.

It wasn't until he was in the bathroom, standing under a showerhead that atomised the water to produce both steam and a waterfall that he relaxed enough to think about how he would advance the plot of his latest novel. Seth had written himself into a corner on Friday and he still had no idea how to fix it. He loathed the thought he might have to scrap what he'd done, as getting this far had already been such a battle.

It would be his fourth novel. One critic had described his previous book, *The Gift of Nobody*, as a "searing exploration of the antiquated notion of privacy in the era of the thingweb." Another critic proclaimed it "had the potential to spark an intense debate on whether it was either safe or wise to allow people to disconnect." But it had seen little in the way of domestic sales — just a few thousand copies — and even fewer sales abroad, so the argument was short lived. None of the major culture aggregator bots had surfaced the online discussions he had participated in, so the book never went viral.

It wasn't the royalties that concerned him — well, not too much, anyway. The basic annual income guaranteed adequate food and housing. What he needed was readers, and lots of them. He was confident enough now as a writer to know his work resonated if he managed to get it into people's hands. And there were so many things he wanted to say, if only he could find more readers.

If only.

His current book, which he'd tentatively titled *Beget*, was about a sculptor obsessed with Michelangelo and who insisted on carving real stone with hand tools, even though he could use a fabber to print something intricate, or a pocket laser engraver and achieve far more precision. Seth wanted to explore what it meant to be an artist when just about anyone could be one, and when the tools required to quickly generate high-quality art in any medium were cheaply and widely available.

He shifted his position to let the water pour down his back. How could any single person be special when everyone was special? And what about things like Tasha? He knew his DPA could produce a great narrative report on the most recent Leafs game by analysing the video feed, and it could write a serviceable romance novel in about twenty minutes if he set the character parameters. Indeed, many people read nothing but the super-formulaic genre fiction churned out by their DPAs. Fortunately, there were a lot more readers — like Seth — who found the characters in these novels flat and lifeless; they were also boring after you'd read a few, much like the sports reports.

Seth stood there for a while, leaning against the shower stall tiles, enjoying the contrasting coolness on his skin. The problem was, if he was honest with himself, he didn't know the answer to any of those questions. And did he truly find the DPA-generated writing boring, or was he just being a snob?

He thought about the impassioned debates back when ebooks had become a big thing. Many people had claimed to be horrified by the very idea of an ebook even though it offered several advantages and the mass print book production era had been incredibly wasteful. These days, only superfans had a hemp or bamboo paper fabbed version of the books they wanted to keep in hard copy, and they were expensive and treasured. His own hard copy library was a thoroughly eclectic mix of authors such as Margaret Atwood, Terry Pratchett, Andre Brink, Michael Ondaatje, Irving Stone, and Maya Angelou.

He sighed, finished his shower, dressed, and walked the short distance to his kitchen to find his food fabber.

"Present breakfast options."

"It is Monday," Tasha replied. "Will you be exercising this morning?"

"Yes," Seth answered. "I have my Kenpo class. But I think I have somewhere else to be today?"

"That's tomorrow," said Tasha. "You have a meal with your family. Home menu choices for that day have been adjusted accordingly."

"Excellent," Seth rubbed his hands together. That would give him the whole day to work. He considered the fabber screen. He had tailored the menu schedule to provide consistent energy throughout the day, but peak energy and alertness in the evening hours, when he felt he was at his most creative. Three meal options appeared on-screen: cricket flour flakes and milk, a termite muffin, and buqadilla, a spicy dish of chickpeas and mealworm protein.

He picked the last option and then had the fabber brew him a cup of yerba mate while he waited for his breakfast to print. "News," he said.

As he sipped the hot drink, its dried grass scent filling his nose, he listened to Tasha report on an explosion downtown. He felt his stomach clench again. A guy might be taken out at any time, by anything, he thought. Even here in Toronto where that sort of thing never happened.

His wristband beeped and he groaned as he looked at it. His grandfather had called out emergency services again; this would be the fourth time in as many weeks. A stubborn man all his life, but worse now that he was in his hundreds, he refused to get a DPA installed or even hire part time human help, preferring instead to call Seth at all hours. And lately he'd started doing things like bringing the paramedics out for mundane stuff like a random blood pressure check. He was going to get in trouble with the police for tying up emergency personnel at this rate.

The fabber dinged and he took his breakfast to his small table, waving a hand over it to cool it faster. He would have to go over to his grandfather's apartment as soon as he'd finished eating, and he'd be lucky to get him settled down and happy again before noon. He loved his Nonno, to be sure, but it was getting increasingly hard to be patient with him. Seth always felt incredibly guilty afterwards if he felt he'd been too short or gruff with the man.

As he drained the last of his drink, Seth offered a quick prayer to the creativity gods to leave him with some energy to make use of whatever time he could rescue later today.

MAURA

Silence descended on the boardroom when the door opened. The staff members seated around the table, her most trusted senior managers, drew in their breath when she walked in.

Maura Torres would have been an imposing figure even without her height and strong build, or her position as head of EduTain. She had dark hair and dusky eyes, which she accentuated with minimal makeup, and a stylish, modern haircut. Her suit was tailored and immaculate, its colour designed to emphasise her olive skin. She walked with an unwavering, measured gait, the hard heels of her shoes striking the floor firmly.

Maura was also very annoyed. She hated unexpected events.

She assumed her place at the head of the table, and her assistant moved quickly to put a steaming cup of tea in front of her. Maura took a few moments to get comfortable and waved her hand over the table surface, bringing a screen to life. Only after she had spent a few minutes reviewing the agenda did she acknowledge everyone else in the room.

"I would like to apologise for being late," she said, pressing her lips into something that might have been a smile on a more congenial person. "Apparently, there was an incident downtown; everything had to be rerouted for security reasons." Maura raised a dark eyebrow and pinned someone with a look. "I hope you've got a report for me. What happened? Is this some local and minor problem, or is there a broader underlying concern we haven't detected yet? Does this have the potential to affect business operations either immediately or in the future?"

Michael, a short and faintly nervous man, straightened in his chair. "I just spoke with the police. At this point, they don't believe it was a terrorist incident. Nothing in their intelligence reports, or indeed, our own, suggests there are any active cells for any group in this city. Rumours suggest it might be something to do with organised crime, but that's pure speculation, in my view, as the local gangs haven't bothered disputing over physical 'turf' in a long time. Everything was automatically rerouted to evacuate the victims. It is still under investigation, but I should get an update in maybe an hour or so. A police spokesperson will make a public statement in about two hours."

"So the assessment is?" Maura prompted.

"In the models I ran using an updated version of our threatcaster analytics software, there is a high probability this was an isolated incident; possibly a maladjusted teenager messing about. It will not have a material impact on our current or future plans," Michael replied.

Maura paused to consider whether she considered the answer satisfactory, and she felt tension draining from her body. The sudden blare of her pod's emergency tone during the morning commute had triggered unpleasant memories, stuff she thought she'd left behind when she'd immigrated to Toronto. She nodded, and Michael quietly exhaled. "Talk to me again when you know more and after you've rerun the models with any new data. Moving on. Danielle," she said, "your report please."

Danielle, who was in charge of business intelligence for the company, looked startled for a moment, clearly not expecting to be called on so soon. Maura was pleased to see she recovered quickly. "Uh, yes ... Our contact inside Sensate has confirmed Project Relax is, as you suspected, their attempt to nail the motion sickness problem once and for all. They are experimenting with transcranial stimulation and supposed calming odours."

"Odours?" Maura raised both eyebrows. "I can't imagine that's meeting with much success."

"No, ma'am. Sometimes, the test subjects are violently ill, depending on the scents used."

Maura nodded. The virtual reality industry had been throwing everything it had at motion sickness research for some years. Hardware improvements had brought latency down and resolution up dramatically, but the human brain was proving stubbornly difficult to deceive. A large portion of the population still experienced nausea when accessing a room-sized VR simulation. Trying to use the sense of smell, Maura thought, was a strange way to attack the problem.

"Interesting," she said aloud. "But rather stupid. That is why I insist all of you know your industry history. Who knows why that applies in this case?"

Michael perked up, looking pleased with himself.

"Smell-o-vision. Something they tried back in the … 1970s? 1960s? Right around then, anyway. Scents released in movie theatres to enhance the immersion experience. Lots of technical problems, and it got a bad reputation straight out of the gate. I don't think it would have gone any further even if it had had a flawless launch, though, if only because scent is a personal experience. Many memories are associated with smell."

"I wonder," asked Danielle, "if they're trying to brute-force predict the associations using data from each individual's DPA?"

Michael let out a low whistle. "That would eat up enormous amounts of processing power, even by today's standards. Especially for older users. All that history to sift through?"

Maura's smile had genuine warmth this time. She enjoyed watching her team interact, particularly during their more thoughtful and creative exchanges. It made her feel like she had family again. She was looking forward to the weekly brainstorming session all the more.

"Yes," Danielle continued. "Crazy amounts. And then there's the issue of using synthetic fragrances."

"Indeed," Maura said. "Allergies are still hugely problematic. As I know very, very well." Everyone smiled. Maura's aversion to colognes and perfumes was legendary. "Okay, encourage our contact to keep a close eye on the transcranial work, but maintain a safe distance from the scent research, or he'll be wanting vomit danger pay. Tell him he has another deposit coming."

"Of course," Danielle replied.

Maura sipped her tea, frowned at it, and set it aside. The assistant paled.

"On to business development. Kwame," Maura said, "what more do we now know about our current acquisition targets?"

Kwame tapped the table and brought up a projection showing the logos of several companies. "I have gone through a dozen scenarios and narrowed the list to these six possibilities, based on financials, staff involved, and technologies of interest. Of these six," he tapped again to change the image into an ordered list, "the top two are optimal."

"Rationale?"

"Xperience has the highly regarded *Dragon Slayer* series which would make the acquisition a PR coup," Kwame said. "Imprint Tech has excellent revenue on the back of its medical training packages but is struggling with cost controls. It's more likely to be receptive to an offer from EduTain."

Maura reviewed the list and made a decision. "Start work on Imprint first. The head of Xperience is self-assured and confident; I met her a few years ago when I got my MBA. She would be harder to negotiate with. But she'll have to consider us if we swallow Imprint as we'll be too big of a threat then." She looked around the table. "I'd like to get back on schedule, so let's save the rest. Until tomorrow." Without waiting for a response, she left.

HAROON

J-District. End of the line.

Seventeen-year-old Haroon got out of his pod, squared his shoulders, and spat. Not for the first time did he wonder why this sprawling cesspit was allowed to exist in the beating heart of the city, surrounded by so much splendour.

He walked around the stacks of crumbling cinder blocks erected at the end of the flow path. Designed to keep pods out, the heaps were more symbolic than functional; they were also covered in rude graffiti.

Haroon stepped off the well-worn footpath and slipped into the shadows cast by the derelict buildings, even though the snowdrifts made it harder going there. Years ago, when the second big real estate bubble had finally burst, and the city began haemorrhaging residents to the rest of the province, Toronto had initiated a great rejuvenation project. The surrounding neighbourhoods had slowly been upgraded. They had flattened old buildings and replaced them with smart, connected structures. Anything of historical value was gutted and refurbished. As very few people bothered to actually own their own transportation anymore, parking lots were ripped up and lush, green parks installed in their place. Many of the roads were reclaimed for pedestrian traffic.

Everywhere except here. Roughly bounded by the 400 Flow to the west, Driftwood to the east, Downsview Dells to the south, and Steeles to the north, the area once known as 'Jane and Finch' had been through several rebrandings, but none of them had taken. Now it was just J-District: a depersonalised name for a region most city residents would rather forget. Squeezed in on all sides by the citywide gentrification, it had become the last refuge for the destitute, the unwell, and the so-called Analogues, a large group of people who flatly refused to stay digital and hook up to the thingweb.

Over time, it had morphed into a big, dark spot on the grid. And with the darkness came the crime. The municipal council had stopped sending the police directly into J, preferring instead to use it as a containment area for the city's worst elements.

It was crowded, it was filthy, and it stank.

It was also Haroon's home.

He felt the simmering resentment again. He'd discovered a book called *Leviathan* recently, while on the outside. Written in sixteen hundred and fifty goddamn one, it had described life *then* as being nasty, brutish, and short. Why wasn't it different by now?

Walking quickly, he hunched his shoulders against the cold and passed by crammed-together shops and rundown tower blocks. Somewhere, someone was cooking dinner; he could smell boiled cabbage. It was cheap and easy to grow, but it always made him feel sick. To him, it was the stench of poverty.

He headed north, crossing the street to avoid the alley where a snarly mutt liked to scrounge through the garbage bins. He checked the sky: the sun was climbing fast and would soon be overhead, eliminating the shadows he was creeping through, making him more visible. The District was not a place where you wanted to be caught walking alone, even in the daytime.

Haroon wondered if he should find a gun. He knew they could be fabbed easily. On the outside, you had to have training and a licence to print one, then you had to purchase the one-time fabber pattern download. He suspected it came embedded with identifying information or some sort of biometric trigger link to make sure only the licensed owner could fire it. And you'd need a silencer to muffle it because otherwise, the citywide sensor network would send a 'shots fired' notification to police. Here in the District? Someone had probably figured out a way to get around all of these controls, but he'd never had the nerve to ask anyone. He didn't want the attention it would bring.

He just wanted out of here.

Behind him, he heard an old motorcycle roar to life, and he winced as the sound blasted up the road, bouncing and echoing off buildings. There were still quite a few stinky gas combustion vehicles in the District, and every time one roared, he tried to imagine how loud and smelly the city must have been when there were millions of them, all growling at once. It boggled his mind to think of that much constant and insistent background noise. They always woke him up here. And emergency sirens! Because apparently, the only way to get human drivers in their noisy vehicles to move was to make even more racket. How had anyone ever slept through the night?

Picking up speed, he rounded a corner and slammed into a huge man dressed in dark clothes. The impact sent him sprawling. Haroon scrambled to his feet, hands raised, and shifted his position to put the wall behind him. "Sorry, sorry. Didn't see you coming."

The man said nothing. He looked Haroon up and down, blinking, assessing.

His fist was solid iron.

The punch crashed into Haroon's stomach and lifted him right off his feet, sending him flying backward. His head thudded into the cold cinder block wall, and he collapsed to the ground, where he lay, writhing in agony, gasping, unable to breathe. He tried to get up on his hands and knees, mouth opening, closing, sucking nothing. The man watched him for a moment, his face expressionless.

Finally, he lifted Haroon and stood him up; still wheezing, Haroon bent double, desperate take in air. In one swift movement, the man grabbed the tail of Haroon's coat to pull it up and over his shoulders, and yanked it inside out and off him. Then he shoved Haroon backward and walked away. He never looked back.

Haroon landed hard on his tailbone, but the pain was a mercy. The jolt was enough to force him to start breathing again. He sat there for several minutes, gulping in deep breaths, clutching his stomach and willing himself not to throw up.

The cold started to bite. Stifling a groan, he got up and half ran, half stumbled home.

His apartment was on the fourth floor. The lift had stopped working years ago, so he had to stagger up the disintegrating stairs. He unlocked his door, banged it shut and let out a long sigh. This was the third time he'd been mugged in as many months and he wasn't sure where he would find another coat, especially in January.

A movement in the next room made him freeze in fear.

His father was home.

RAY

The drone had exploded. That much he knew.

Blood. There was blood everywhere.

There were people moving, talking, shouting, hurrying.

They were so close, but they all sounded so very far away.

The lights. So bright. Halos around them.

Ray shifted, trying to see, and then screamed — a long, horrible wheeze-shriek that was alien, weirdly muffled. His body was on fire, filled with hot pockets of coal burning him from the inside out.

Air… He couldn't get enough air.

"Blast trauma," someone said. "Some sort of explosive device. Had already gone into hypovolemic shock before the parameds injected the wound stasis foam. Get more synthblood racked up."

"Mobile scans showed a ruptured tympanic membrane, collapsed left lung, probable alveolar haemorrhage, and an alveolar rupture in the right," another man said. "Secondary damage includes thirty-four penetration injuries and counting. Splenic laceration and rupture, multiple fractures. Head trauma from where he was thrown by the blast."

"Why haven't we determined our lockdown status?" asked the first man.

"On-site scans were clean. Looks like we've got shrapnel from the device, bone fragments from the other victim, his own bone fragments, brick fragments, bits of his clothes and multiple empty sites that were probably ice shards, but no biological agents. This seems to be an old-fashioned bomb."

It was a drone, Ray thought. Not a bomb. But he couldn't work his mouth to say the words.

"I don't like the sound of 'probably.' Get the high-resolution equipment in here. I sure don't want to be the last to find out some jackass with a basement kit has managed to hack a new henipavirus and we're ground zero."

"On it."

A woman leaned over Ray and looked at a display above his head. She was dressed in brightly coloured scrubs with a strange, dancing pattern on the tunic.

"When are we getting the RCMP forensic data share?" she asked, over her shoulder. "I want to run the projections on other injuries and outcomes."

"They're on scene now with the Toronto police," said the second man. "Maybe thirty minutes before they upload?"

Where was Mick? Ray wanted to ask the woman. Mick had always been there before, to help him, protect him. Mick was his best friend.

"Tabarnak," said a new voice with a strong Quebecois accent. "This guy? *Un drôle de moineau*. He's got none of the implants, no augments, no biomems. A tattoo, does he have one of those?"

"Can't tell," said the woman. "That arm was shredded."

Someone else was slicing open his tattered and smouldering

clothing, peeling it off section by section, teasing it out of his wounds. Every movement brought another wave of excruciating pain that made him convulse.

"Turn it up, get this poor bastard to lights out."

Ice filled his veins.

Blackness.

KEL

Out in the hab, Kel examined the macaque's body. "Meike, what do you suppose happened here?"

Meike Bergholtz, her assistant, was standing nearby, leaning heavily on the upright stretcher they would use to retrieve the macaque. She had a strange, almost severe face, with eyes set just too far apart, and a thin, pinched nose. She wore her hair clipped short at the back and on the sides, keeping it longer on the top; Kel thought it did nothing to soften her appearance. Meike shrugged. "Eh, probably a dominance fight." She wiped the sweat from her forehead and yawned.

Kel crouched and suppressed the desire to glare at Meike. "Try looking, will you? I don't see any bites or gashes." She swatted away the flies that were already gathering and gently rolled the macaque's body over. The macaque's head lolled to the side at an awful angle. She felt under the fur on the back of his head. "Implant's intact, at least."

She looked up to see Meike picking at her fingernails. She sighed. "Leave the stretcher here and go find Dalton. Keep your tranq gun out as he may still be alive and in pain. He'll bite."

Meike let the stretcher fall to the ground and took out a

small sidearm. She wandered away.

Kel reached over Max's body to pull the stretcher over and then pulled him onto it. She arranged his arms and legs, closed his eyes, and stroked his brow in the way that once would have made him burrow in close for a cuddle. She sniffled a little and stood up to survey the area.

She was at the edge of a clearing, under an enormous mangosteen tree. Kel inspected the ground where they'd found the body. The soil was hard here, held in place by the network of roots created by the plants covering the forest floor. There was no obvious damage to the surrounding undergrowth that would suggest there had been a fight, or even a trail to indicate there had been a chase. Kel glanced up at the tree. It wasn't unheard of for a macaque to fall out of a tree, but they were usually surefooted.

Meike sauntered back. "Found him," she said.

Kel waited a moment, but nothing more seemed forthcoming. "And?"

"He's dead."

Kel bit back an exasperated reply. She pointed at the stretcher handle. "Lead the way."

Meike picked up her end and they moved further into the trees.

The macaque was a crumpled heap. Kel unfolded his body and looked him over, smoothing out his fur. He was still warm, and she couldn't see any gashes or bite marks. There weren't any signs of conflict on the ground here, either. He also had a broken neck.

Kel rubbed a hand up the side of her face, puzzled. Two

macaques dead on the same day, both with the same injury. What were the odds?

They put Dalton on the stretcher before they trudged back to the office, passing through it to get to the lab. Kel wiped her cheeks with the back of a sleeve.

"Are you alright?" she asked Meike. "Do you feel you could autopsy them?"

"Of course," Meike looked at her strangely. "Why wouldn't I?"

"Well, it's just that you worked with them clo—" Kel began, but stopped when she noticed Meike's odd, expressionless gaze. "Okay, fine, it's never been my favourite thing to do. I get too fond of them. Recover the implants and do a forensic DNA analysis."

"Obviously."

Kel gave up trying to connect with Meike and retreated to the main office. She was reviewing the overnight logs when her boss, Robert McGee, strolled in. A heavy man with a barrel chest, Robert's clothes were perpetually rumpled, and his hair was never quite in place. He pulled up short when he spotted her, clearly not pleased to discover he was not the first one in.

"Did you even go home last night?" he muttered, heading straight for the office fabber and ordering a coffee.

"Morning. Yes."

"What time?"

Kel looked blank. "Good question. I left when I got too tired to see properly."

"Hmph." Robert slurped at his coffee. "What are you doing now?"

"Uh, checking the overnight summaries."

"Well, I've got budgeting work to do," he announced, as though Kel had been holding him up. "So I need to get on with that."

He went to find his desk, leaving Kel to wonder why she had hedged with him just now. Best to leave the bad news until she had some answers, she thought. She returned to the logs to see if she could uncover any clue why two of her macaques had died so suddenly.

MAURA

Maura's office was huge.

Located on the corner of the top floor of a tower on Queen's Quay, it offered commanding views of Lake Ontario and the ferry terminal on one side, with the city sprawling to the southwest on the other. She kept the furnishings to a minimum here. Her desk sat at the apex of the panoramic windows and faced into the room; two comfortable chairs had been positioned in front of it for private conversations and interviews. There was a medium-sized worktable in the adjacent corner, while the corner opposite to the desk was a large section that could be partitioned off to create a VR demonstration area.

The walls that weren't given over to windows were reserved for her favourite works of art: a big print of an orchard by Van Gogh, a beautiful still life by Henriette Knip, another Dutch painter, and a dramatic northern landscape by Lawren Harris. The only other artwork in the room was on her desk, a surrealist sculpture called *Bitoro*, by Francisco Pereira, which looked like a bull reimagined as a long-legged biped. Maura often stared at it when she was deep in thought; it reminded her of the absurdity of the world and helped her see things from a different perspective.

At that moment, however, she was watching the screen set in her desk. It displayed the view from the camera in the lift that brought people to her office. The lift was gliding to a halt and the young woman inside it — Pauline McDonald was her name? — touched the control panel twice to change it to 'doors hold.' Maura watched as she took a few seconds to take a deep, steadying breath. The woman flicked a quick glance at her reflection. Her dark suit was spotless, her shoes classically styled and tasteful. Her jewellery was simple — accenting rather than overwhelming her outfit. Her long, blonde hair was swept into a basic Gibson tuck. Her expression was calm and resolute. Maura approved.

Pauline tapped the panel again to open the doors and walked into the room, shoulders back, head held high. She stopped a few feet from Maura's desk and waited.

Maura did not glance up from her screen, which now displayed the projections for the next fiscal quarter. A minute stretched into two, and then into five, and then into ten.

Pauline stayed as still as could be.

After several more minutes, Maura looked up. "Tea," she said and went back to reading.

It was a test, of course. A menial task after being forced to cool your heels for what must have felt like forever. Although her company had a gruelling interview process — three separate sessions, including one in front of a panel — and used all the usual psychometric assessments, Maura still liked to throw things at her prospective assistants to see how they reacted in real time.

Pauline scanned the room, spotted the food fabber near the window. Briskly, she walked towards it, finding there was a storage cupboard to the right. She opened it to reveal a single, simple place setting. She removed a teacup and saucer, placed them in the fabber, and swiped her hand along the control screen. Her fingers flew across the menu, and a fragrant steam billowed upward. Tea, green, Pi Lo Chun.

Maura nodded to herself, impressed. The woman had done her research.

Pauline brought the tea back to Maura's desk and wordlessly set it down. Without looking, Maura reached for the cup and took a sip. She put it back in the saucer. She made sure her expression betrayed nothing.

More waiting.

At last, Maura turned off her screen and gave Pauline her full attention.

"Have they told you I'm hard on my assistants?"

Pauline hesitated for only a second. "Actually, the exact phrase was 'chewed up and spat out her last four.'"

Maura paused in mid-reach for her tea. She arched an eyebrow. "Oh? Would you care to inform me who said that?"

"No."

Her eyebrow went higher.

"If I'm to be effective, I have to tell you how things are," Pauline explained. "But I also have to get along with the other members of your team."

"I see," Maura replied, and returned to drinking her tea. She sipped it pensively. "Why?"

"Why do I want to work with you?" Pauline ventured.

"*For* me," Maura said sharply and put down her cup. "The hours are ridiculous, I'm particular and demanding, and little of the work would seem meaningful or fulfilling. Much of it, as you must know, is comprised of things a DPA could do. I prefer having a human assistant here at the office. I am always curious why anyone would sign up for this."

Pauline nodded. "You're known for doing what you do very, very well. You personify the pursuit of excellence. You are hard, but fair, because you demand that people are their best selves when around you. And in the absence of the grander purposes our predecessors were once offered, I feel there is much to be learnt from your approach to life."

Maura permitted a flicker of surprise to dance across her face.

"But more than that, by my reckoning, your company is now fifth in the alternate-realities industry worldwide. I'm certain that with the proper personal assistance, EduTain can get to the first position. I'm also sure I can provide you with that support."

"Working for me as a higher calling," Maura said wryly. "That's a new one." She considered Pauline for a few more minutes, amused, and again intrigued by the woman's poise. "See David in HR about integration," she said and returned to her screen to get back to work.

SETH

Seth stepped out of his pod across the flow from Il Contadino and hesitated, feeling slightly hopeful. Maybe it was the wrong day? As much as he adored his family, just the thought of spending the whole evening with all of them at once was draining. Then he saw his cousin Joe — Joseph Bacchi, Queen's Counsel, as he preferred to be called these days — get out of another pod nearby and head into the restaurant. He smiled ruefully. There would be no escape.

Inside the restaurant, he groaned at its antiquated setup. They still employed human servers here! He tried very hard not to think about germs and potentially fatal diseases.

The maître d', a plump, dark-haired woman, approached. "Hello, I'm Ambra. How can I help you?"

"I'm with the Bacchi family luncheon," Seth replied. "We're in the Ribiero room, I believe."

"That's all one *family* back there?" she said, forgetting herself momentarily and looking astonished. "I thought it was a business dinner."

"Yep," Seth put on a grin. "I'm one of seven kids. My father's brother has six children. Italian on one side, African-Canadian on the other. I could go on."

"Wow." Ambra gestured for Seth to follow. "You rarely

see that size of family anymore."

"No, no, you don't," Seth muttered. He had often wondered why all of his relatives seem compelled to have so many children, especially when it seemed like everyone else in the world was doing the opposite — even countries formerly bursting at the seams, like China and India, had seen their populations decline. Some of it was down to Bacchi tradition, to be sure, although no one in his clan had been anything but nominally Catholic in more than two or three generations. He'd asked his mother once why she and his father had signed up for so much childcare, not to mention all the drama that came with a big family. She just laughed and said they loved it.

Privately, and on his more introverted days, Seth thought it was all just a bit selfish to foist so many kids on the planet.

Ambra led him to a room at the other end of the restaurant. When she opened the door, the noise rolled out like a tidal wave. He went in and felt, rather than heard, the door close behind him.

It was a typical faux old-world Italy dining hall. The walls were covered with garish Renaissance-era murals in the style of Mantegna. There were the requisite white pillars, and the chandeliers were built to look like bunches of grapes, which made Seth roll his eyes. He preferred the gatherings that celebrated their African roots, as those establishments seemed so much more authentic, at least in this city.

The restaurant staff had lined several tables up in rows to accommodate the crowd, and the room was already warm. He found a chair that looked like it wasn't claimed and slipped into it. Someone came by, filled his wine glass without asking if he wanted any, or even what he wanted. He debated whether he should have some this early in the day —white wine was not his favourite — before he finally gave in and took a sip. He figured he would need the alcohol to get through the afternoon.

Seth settled back into his chair and into observer mode. There was his Aunt Sandy, the quadcopter adventure guide based in Northern Ontario, who talked not just with her hands but her whole body. She was clearly telling a tale about one of her children. Cousin Jember, who oversaw the city's waste management program and had a tendency to slouch, stood by his father, hands in his pockets, nodding occasionally. Two youngsters — were those Abele's kids? — had commandeered some butter knives and were having a sword fight in the middle of the room.

He had drunk nearly three-quarters of his glass before someone — his mother — finally noticed him and landed on the chair next to him. "Seth!" she exclaimed and kissed him on the cheek. Her eyes sparkled. Family gatherings always buoyed her. "*Cucciolo*. I didn't see you come in."

"Hello, Mamma," he replied, setting down his drink so he could take her hands in his. "How are you?"

"You really need to come by more often," she said. "What have you been doing that's keeping you so busy?"

"Well, I've been working on…" He stopped as his view of Mamma was blocked by the hulking mass of his brother Dario, who had swooped down to kiss her.

"Dario, sweetheart, so glad you could come this time!"

She patted his arm and looked at Seth. "Did he tell you he missed our last dinner because he was in Zimbabwe?"

"No, I hadn't heard," Seth replied. "I've been—"

Dario nodded. "Expanded the research facilities there. The technology is moving so fast right now. We're close to regenerating skin right on the body instead of fabbing it and grafting it."

"What? Oh my, how exciting," Mamma said.

Dario absentmindedly picked up Seth's glass and drained it, shaking his head. "We've been able to do this kind of tissue engineering for a while, originally using umbilical cord stem cells or even cells from the gums. It's got hair and functioning sweat glands. But grafting is painful. Controlling skin regeneration has been the big problem to get around. We think we're close. I've been busy trying to poach talent from Tokyo these last couple of weeks."

Still holding Dario's arm, Mamma said, "That's amazing. Imagine what it could do for burn victims, like those poor souls in the building fire in the District last month. But I would guess no one wants weird, excess skin growth from a treatment."

"Exactly," Dario said. He glanced at Seth. "Life good at Reprint Tech?"

"Uh, Imprint Tech, I think it's called," Seth said. "But I don't work there —"

Dario nodded. "Right, right. The medical training company. They produce good stuff. Say, is okay if we drop off the kids on Thursday instead of Saturday? My boss has ramped up my schedule like you wouldn't believe, and I want to fly out early to get a handle on it."

"Well having them for the weekend is still fine, but I'm trying to —"

But Dario had already turned back to Mamma. "So I hear Joe might have someone new?"

Seth gave up. It was always like this. It didn't matter which one of them he was talking to. He started fretting as to how he was going to stretch the two days of activities he'd planned for his nephews into four days.

"Oh, I couldn't comment," Mamma said. "He didn't bring her tonight, so I can't tell how serious it is."

"Sandy expecting again?"

"You know, she isn't. She's thinking that will probably be it for her," Mamma replied. "Just the four kids."

There was a loud crash, and then a wail. Seth turned around. The sword fight had, predictably, gotten out of hand.

"Oh my," Mamma said, rising. "I'd better go give out nonna kisses. Dario, be a dear and start shooing people into their seats, won't you? I expect they're ready to begin feeding us."

"Of course." Dario thumped Seth on his back, nodding curtly before walking away.

Moments later, a server approached. "Wine?"

Seth picked up the empty glass and handed it to her. But before he could say what he wanted, she refilled it with white. He fell back in his chair, wondering if anyone would ever think to ask him first.

RAY

Drifting in and out of consciousness. A memory.

He was four again.

He was… he was standing on his bed, goggle-eyed with terror. His knees wobbled so much he could barely stay upright.

In the other room, he heard his mother, screaming, spitting with rage. Ray didn't know why she was angry this time.

Warm wetness slid down his leg, pooling at his feet before soaking into the tattered mattress.

He didn't want another beating. He would have to hide. Now.

But the *thing* was back, hissing at him from under the bed. He was sure of it. It wasn't in his head, like she said. It waited for him, in the dark.

Hot tears poured down his face, and he gulped and whimpered.

But there was nothing else in his room. Just the bed, the grimy walls, and a window that always seemed covered with frost on the inside. Nowhere to hide.

Something smashed against his door. He flung himself backward, landing hard, twisting an ankle. He turned over and crawled headfirst under the bed, swinging his fists in a desperate attempt to fend off the monster. Then he pressed himself into a ball and buried his nose into his sleeve, willing himself not to sniffle and give his hiding place away.

~

Consciousness… strong lights… gummy, sticky eyelids that wouldn't open very well.

A nurse, noticing the fluttering, came over and smiled. "Whoa, easy now," he said. "You can relax. You're out of your first round of surgery and in ICU. Lots more to come, but we have to do this in stages. Not good to keep you full of the anaesthetic all the time. Just rest. Are you in pain?"

Ray couldn't answer. It was all pain.

~

His eleventh birthday.

A blistering hot day, with the sun so strong he felt the tops of his ears cooking.

The knife made a noise almost like a zipper as it tore down his forearm, opening his flesh.

Ray staggered back and blinked. The pain came a moment later, sizzling down from his elbow to his wrist, and the air stunk like overheated metal.

"You little *hūndàn*!" the man spat. "What you looking at?"

The man was short and hunched over. His left eye looked off to the side while his right eye bored a hole through Ray's head. He had several black teeth.

"I said what you looking at!"

"Nothing! I…" Ray looked down and swayed. His arm was dripping already.

A hunk of concrete whistled past his ear and smashed into the man's face. The man reeled backward and down, hitting the broken pavement hard, the knife clattering to the ground.

Mick came up from behind him, grabbed his other arm, and pulled. "Run!"

Running. Running. So much running. Tumbling down the bank to the drainage ditch that ran under Finch and pushing away the overgrowth to get into the hole between the concrete and the culvert. Mick's sanctuary, a secret hiding spot. Mick was always saving him. Mick was everything.

Blood. Cheap scotch poured all over his arm. Fire and flames that seemed to go all the way up, into his chest, and down to his gut. The smell of Mick's sweaty shirt, torn into long strips and wrapped around his arm.

~

Somehow, he came to during a surgery. He was draped in sheets. A massive surgical robot hovered over him like a steel octopus, two arms snipping and probing, another pair poised with a light and a tiny camera. He flicked his eyes left and saw a blurry rectangle on the wall; he blinked, and the image resolved into a high-definition screen. There, in full colour, were the shattered remains of his gut.

He could hear beeping over his own gasping into the tight mask over his mouth and nose. His breath was hot.

Suddenly, his face and neck were blazing, he couldn't breathe, and he felt like he was drowning, underwater, breathing hard, the beeping went faster, he couldn't move…

And then he melted into the darkness.

MEIKE

In the lab, the macaque's body was laid out on the table, now covered in black dots laid out in a grid pattern. Meike turned to a touch screen and used a gloved finger to poke an icon. Overhead, a camera moved silently on a track and imaged the whole body.

When it was finished, she flipped the creature so the macaque was face down, and set to work putting more dots all over the macaque. When she was done, she put the camera on again. She stared vacantly at the readout screen while it ran through another imaging cycle.

This felt like her childhood all over again: nobody around, and nothing but a variety of screens — some big and wall-mounted, others handheld — to babysit her, and feed her media. Flat images, flat people, living vicariously through endless videos, feeling nothing.

A three-dimensional image of the outside of the dead macaque rendered on the screen. She saved it to a file, removed all the dots, and scooped up the body roughly in her arms. She took it over to a waiting body bag and zipped it in; the other macaque was already in a bag of its own. She slid both bags onto a gurney and wheeled them out of the lab's back door, which opened onto the facility's quad. The air was bitterly cold, and there was enough of a breeze to make it gnaw at Meike's cheeks. Even though she didn't have her coat, she walked at her usual slow pace across the open area. The pain on her face, the stinging, the burning, was exquisite.

By the time she reached the other side, she was shivering hard. Opening the door and stepping into the warmth was a disappointment. It was safe. Normal.

Flat.

The other sections of the quad consisted of a state-of-the-art veterinary hospital they shared with the zoo. She eventually arrived at the imaging and diagnostics department. It contained several different scanners arranged in a series, each one connected to another by a short section of conveyor belt. As most live patients typically objected to being passed through a group of scans without human intervention and reassurance, this setup was reserved for examining dead bodies and inanimate objects.

Meike unzipped the bags and dumped the bodies out onto the cold trays that would carry them through the scanners. She ordered a thorough diagnostic set and then walked to the other side of the section where the macaques would roll out in about an hour.

There was a waiting area here, complete with a few chairs and a small food fabber. Meike sat on the chair and fidgeted. She stared at the grey machines, the grey walls, and the grey ceiling. The lights flickered briefly, as the first scan began. She focused on the light directly overhead until her eyes ached and watered so much she couldn't keep them open.

She was still wearing her gloves. Meike pulled the stretchy cuff of one as far as it could go and then let it snap back into place. She did this again and again, and then again. Her wrist reddened, but it didn't feel nearly as interesting as the wind had done.

On the table near the fabber, she spotted two cylinders filled with forks and knives. On a whim, she jumped up and grabbed a knife. She felt the hardness of the handle in her right hand, the weight of it balanced against the silver blade. Suddenly, she could smell the rubber of her gloves, the scent sticking in her throat. She splayed her left hand, and drew the tip of the knife across the upper palm, just below her fingers. The knife carved a long line across the taut rubber, slowly, slowly, and then in a flash, the glove split, parting to reveal the pink flesh below. She pressed harder and watched as the skin split too, red blood welling up and seeping darkly underneath the glove. The searing pain raced up her arm, and she sucked in a sharp breath.

The door to the department banged open. A technician walked in and checked the manifest screen before noticing Meike. He strolled over for some friendly conversation, but his smile quickly disappeared when he saw her hand.

"Oh, hey, wow, are you okay?" He quickened his pace to get to a first-aid station on the wall. With his back turned, Meike put the knife down on the table. "I keep telling them to restock the thing with the blades down," he said. "I poked myself only last week. We might have to raise this with the health and safety committee."

He brought over a small, portable haemostat and an

alcohol swab. "Here, let me help you." He peeled off her glove, carefully dabbed the wet swab on the wound to disinfect it and ran the device over the cut.

Meike watched as the bleeding stopped, felt the skin tighten and pull back into place. In a moment, there was little more than a reddish line.

"There, that should do it. You're lucky that wasn't deeper or longer, we'd have to take you to emergency."

Her hand tingled and stung. She thought again of the cold outside and hoped the scans would hurry.

KEL

A few days later, and another late night had morphed into early morning. Kel slapped the top of her desk in exasperation and pushed her chair back.

"Wake screens," she said.

The wall shimmered as the various camera views came into focus. The macaques were just stirring. She put her feet up, thumping her heels hard on the desk, and crossed her arms, glaring at a habitat that had stubbornly refused to give her any clues as to what had happened.

The refusal was all the more galling because of the wealth of data Kel had at her fingertips. The habitat, besides being a faithfully reproduced biome, also contained a vast network of hidden sensors. There were the cameras and microphones of course, but also devices for measuring everything about the habitat like temperature, air quality, humidity, airflow, and everything about the residents, like movement, speed, thermal and spectral signatures. And her implants reported many kinds of biological data, from heart rate to bacterial load. It was a micro thingweb all by itself.

And yet with all of that, there was nothing to explain why two of her subjects had just dropped dead. The computer hadn't found anything, and neither had she.

Robert trundled into the office, and she straightened up quickly. As usual, he went straight for his coffee, but instead of heading into his own work area after that, he came and sat on her desk.

"So," he said, slurping. "What's going on?"

"Nothing much, why?"

"Oh really?" Robert looked annoyed.

Kel braced herself. She drew a breath to tell him about the macaques.

"What happened to this quarter's research paper?"

Kel groaned. She'd forgotten all about the deadline. She was supposed to produce a minimum of four per year, and posters as well. At least one paper had to be accepted and published in a major journal in her field.

"Yes, *that* paper," he said.

"I'm sorry, I have a draft," she was astonished to hear herself lying, "I just need to polish and submit it."

"I have a hard time defending your project as it is, Kel," Robert said. "You've got to produce something regularly. This habitat eats up heaps of public money, and your current study isn't necessary."

Kel was shocked. The display on her desk chose that moment to change images, and a much older version of her grandmother, looking dull-eyed and slightly sad, stared out at her. The image gave fuel to her indignation. "How can you say that? We have thousands of people diagnosed with Alzheimer's every year!"

Robert waved a hand dismissively. "It's thousands, not tens of thousands, these days. And drugs manage the symptoms quite well."

"They still come with side effects and cost a lot!" Kel said hotly. She couldn't believe he could be so cavalier. "Why wouldn't we want to understand it better and stop it from happening in the first place?"

Robert pinched the bridge of his nose. "This is what comes of having to give hotshot kids access to the big toys straight out of grad school," he muttered. "Did you not spend any time in the industry? Or socialising with your profs in person? Grab a beer with anyone after a conference?"

Kel raised her chin a notch. A smart, driven woman, she'd fast-tracked through most of her studies and achieved her PhD at University College London by age twenty-two. She'd studied hard and had spent most of her nights doing lab work. The conferences Kel had attended because of her degree requirements had felt like a nuisance. Why waste time travelling to these things when you could just read the proceedings after and do a quick VR discussion with someone you wanted to follow up with? Until now, she hadn't thought her approach was a liability.

Robert tried to smooth his hair. "Fine, let me be the first to disillusion you: the reason we're publicly funded is private pharmaceutical manufacturers have never been very interested in curing diseases. Drug research is incredibly expensive and really risky. Drugs for disease management are income streams for them, so that's what they focus on. If they cured people, how would they recoup their R&D costs, much less make a profit? I'll bet all your job offers were from publicly funded places, right? None from private corporations?"

Kel frowned and nodded at him.

"And as for the public, ever since we got really good at microfluidics for targeted drug delivery, stuff like this has dropped out of the 'I need to care' file. As long as Grandpa seems okay and isn't wandering into the flows at night, it solves the problem. I can't believe someone with your brains hasn't cottoned on to this before."

"These people are still suffering!"

Robert nodded. "But they're hurting much *less*. And living longer. And those drugs cost a lot less than they used to, right? Frankly, there are other, bigger issues taking up more news cycles. Now if you were working on antibiotic resistance, maybe finding us some novel antibiotics, that'd be a different story." He paused for another long slurp. "We'd be swimming in funding from all kinds of sources. But you're not, and thus we're not, so get those papers done on time. You must produce research data that justifies the resources you use up here."

Her colleagues were filtering into the office. Robert left her to go talk to Padraig, the resident entomologist, and Bao-Yu, a visiting herpetologist.

Kel leaned forward against her desk, exhausted. More than anything, she wanted to go home and get some sleep. But even with her track record of putting in long hours, she suspected leaving just now would not go over well.

Nothing for it but to get something written, she thought. Did she even have anything useful right now? Feeling desperate, she pulled up the implant log for Pika, a newborn macaque almost certain to become diseased later in life. It would be a stretch; but she consoled herself with the thought it wouldn't be the only academic manuscript produced just for the sake of getting a paper published. Kel had once dated a maths postgrad student who had spent much of their first — and last — date telling her how he'd churned out several papers by running the same Monte Carlo simulation with different starting parameters. Yes, she could look to see if there was any interesting preliminary data or any way to compare Pika's baseline to others in the database with a narrative that would be meaningful enough to produce a paper.

Only there wasn't anything interesting. In fact, there was no data at all.

The entire log had been erased.

MAURA

In the summer, one of Maura's favourite office break-time things to do was watch the meticulous agribots tending their open-air rooftop food gardens, carefully weeding, or pruning, or vaporising pests. Their work was so orderly and so precise.

Maura sometimes thought of Toronto as a living body, with the flows being the streams of air and life-giving blood and all the little bots and people like cells, all so different, all equally important, and each with their individual jobs to do. And the thingweb was its nervous system: each tiny sensor a neuron, measuring and reporting on things like load bearing stress on the bridges, air quality in the suburb, and noise levels in the downtown.

She also loved to watch the vertical wall gardens, the lungs of the city, filled with various leafy plants and ornamental grasses that helped to filter the air, as they waved and flittered in the breeze. Their random movements stood in sharp contrast to the more rhythmic tick-tick-ticking of the microturbines or the not quite natural flashing of the leaves on the piezo-trees that dotted Toronto and provided electrical power. So much quiet activity. It was all so calming.

Today however, the roofs were smothered in snow, and the walls had been covered to protect the dormant plants there, so she had turned her attention to the enormous lake that Toronto overlooked. The wind always blew hard across the vast expanse of water. If the conditions were right, wind and the water motion could break up the ice cover and fling large sheets of ice towards the shore. Over time, the slabs would stack in a haphazard fashion. The piles were already taller than most men were; it looked as though a giant had smashed the windowpanes of his castle, leaving huge shards in jagged heaps. Maura admired the raw power nature still wielded over the landscape.

After a few minutes of quiet contemplation, she moved to the other window and glanced down at the city traffic flows, below. "Overlay," she said to the glass. "Pod occupants." Immediately, all the pods sprouted labels showing who was inside; data hacked from the various public and private networks and piped into Maura's personal server. The majority of the labels were white, but a few people that particularly interested Maura were highlighted in green.

"Pauline," she said quietly. The sensors in the room heard the request and pinged Pauline in her office.

A few moments later, Pauline came in. She walked to the window to stand next to Maura. After a moment, Maura said, "Did you know this city produces nearly thirty percent of its own food?"

The question seemed to surprise Pauline. "I didn't. I assumed we brought most of it in from the countryside."

Maura shook her head. "We bring in most of the naturals — the plants that create large amounts of biomass — from outside the city. But all the feedstocks, like the dwarf grain and insect flours used to supply the food fabbers, are grown within city limits. Rooftop gardens like that one over there produce about ten percent. The other stuff, such as the lightweight, fast-growing leafy greens, is grown in the vertical farms like the ones in the re-purposed condos east of here. Just one of them produces about eighty thousand heads of lettuce a day."

"I didn't eat any naturals until I was in my twenties," Pauline said. "All we could afford were the basic fabber prints, one set for every day of the week. If it was Monday, you knew what was for dinner. Flavoured pasta, in our case." She paused, looking thoughtful. "I think the first thing I tried was a cherry. Even now, I have a difficult time getting past the idea I am eating part of an actual tree."

Maura smiled faintly but said nothing.

"You know a lot about the city," Pauline added.

"I make it my business to know a lot of things about a lot of things, whether or not they seem immediately relevant." Maura gave Pauline a look meant to imply Pauline should do the same thing. "Now then, we have a meeting with a certain Councillor Brown in twenty minutes." One of the green pod labels on the window brightened: Brown was on his way. "Tell me what you know."

Pauline nodded. "Brown wants to become mayor in next autumn's election. He's sixty-two, married, and became a father for the second time earlier this year. He entered politics about five years ago, gaining his council seat after a viciously personal campaign, and a platform against crime, immigration, and taxes. They're still piecing together his new platform, but there are hints he plans to do the same again, only with more venom. He's coming to us to secure good rates on an advertising buy in our flagship entertainment bundle."

Maura walked back to her desk. "All of that is in the public domain. What did you learn in our dossier on him?"

Pauline followed her. "He is not faithful to his spouse. He has two steady relationships outside his marriage and he's trying to start a third. Brown's seen his new daughter twice to date, as he is attending functions at various locations around the region to raise his profile. He hasn't paid for most the recipes in his food fabber; when he's travelling, he spends his time accessing pirated VR material through a series of four or five proxies. Including our stuff, of course."

Maura pressed her lips together and sat down. She decided to probe Pauline a little. "Of course. What do you think of his platform, Pauline? Should EduTain support him or his rival for leader?"

"It's not for me to decide," Pauline demurred, "and it's too early to tell. Normally, I'd vote for his opponent, as I like most of her platform, especially the upgrades she wants for the digital voting system, but she doesn't excite me much. And there are things in her past—"

"So because the other candidate isn't perfect, you'd consider voting for someone with an open track record for promoting fear and hate?"

Pauline looked startled. "Well, not necessarily. Uh, as I said, it's early. What he plans to *do* about, say, crime is not clear."

Maura leaned back and folded her hands in her lap. "Ah,

but it is. This morning I collated recordings of every one of his public pronouncements. I ran a sentiment analysis and extrapolation algorithm. It is projecting his platform will be to deport every single criminal, and all residents of J-District to Base 53, and somehow, this miraculously solves the city's problems."

"But that's… absurd," Pauline said, struggling for adequate words. "The International Lunar Port Authority would never agree. It would violate at least six treaties and two conventions. And the logistics of moving how many people out in a short period? And the expense! Surely it would blow the budget?"

"Of course it's ridiculous. But a big enough lie told boldly enough…" Maura replied. "Listen, it doesn't matter if what the other side proposes is total nonsense, or it's beset with scandal or incompetence. Voters won't care if your side has rock solid policies backed by data that benefit everyone. Elections are won by the team that picks a simple message that resonates with the majority of people who bother to vote, and repeats it every single day of the campaign."

"So you think he will tap into some latent fear of people in the District?" Pauline asked.

"There are a surprising number of people who automatically distrust anyone who is not hooked up to the thingweb. It's assumed that if you're not connected, it must be because you have something to hide. It will get worse if he continues to lay the groundwork as he has been doing so far. Not to mention the legitimate concerns about the growing resurgence of organised crime, there and elsewhere in the city."

Maura brushed a speck of lint from the front of her jacket. "If he gets elected, my bet is he'll 'discover' aspects of the law that will stop him from fulfilling his promise. But in the meantime, he will foment unrest, and people will get hurt. I won't have that. J-District is a problem area that needs help. We can't let someone use it to upset the order in the city. Or anywhere else, as I'm sure he has aspirations outside Toronto."

Pauline shifted her weight from one foot to the other. "I wonder what he'll make of the incident downtown, the bombing or whatever that was. Probably nothing good. So what will his real agenda be?"

"I can't tell. Beyond just getting elected, and the usual personal enrichment, of course. His communicator is too well encrypted. I haven't broken it. Yet."

"Then we won't be accepting his proposal today?"

Maura smiled. "Oh, I'll take his money. He's set up an impressive campaign account, and he's likely to spend lavishly if he thinks he's getting a good deal on an influential media package like the VR we produce. And my war chest needs filling."

"Then…?"

"Then our job today is to keep him fixated on bargaining for placement in our top programs. And… not so focused on firming up details on how we deploy his spots. Once he's locked into a contract, we have a range of options for influencing how he's perceived." Maura permitted herself a small chuckle. "One tactic that was effective in the last election was to digitally copy and paste the latest populist idiot's eyes onto the villain of our Dark Wings series. Our demographic analysis told us the people most likely to vote for him were also keen players of that VR sequence." Maura remembered her amusement in reading the confused messages between campaign managers when the polls plummeted and no one could figure out why public opinion had shifted so radically. Even the press, which been almost fawning in their coverage of the 'rogue' candidate, turned on him and ripped him to pieces.

Pauline looked shocked. "But isn't that—"

Maura cut her off with an impatient gesture. "I despise demagogues. We've suffered through more than enough of them through history, and I fail to see why we should again if we have the means to stop them before they get started." Maura's gaze rested briefly on the sculpture on her desk. She cleared her throat. "Progressives are always too pure to take the necessary steps to prevent a backslide. We think well-reasoned arguments and nuance will somehow carry the day. Or we're shocked to learn after the fact that the other side wasn't playing by the rules. If no one else is prepared to nip this in the bud, then I am."

"I see. Then perhaps you should also know he ran an ultramarathon yesterday, and so he will be tired."

"Did he now?" Maura feigned surprise. That the meeting was to take place today had not been a coincidence: Maura had chosen the date for exactly that reason. She waited to hear what Pauline would say next.

"Yes. Oh, before we get started, is there anything you need? I forgot to mention the lift is scheduled for maintenance." She glanced at the window where Brown's pod had nearly dropped off the overlay as it was so close to the building. "In about ten minutes' time. So he must take the stairs. Can't be helped."

Maura wasn't disappointed. The meeting would be even more interesting than she thought. And so was Pauline.

HAROON

"*Yo, tomodachi*," Yoshi said. "Are you sure you don't want me to fab you something for your eye?"

Haroon looked up from his screen to smile at his friend and then tried not to flinch. The area from his cheekbone all the way back to his ear was still sore. He switched from smiling to shaking his head. "No, Yoshi, thank you, really. You do too much for me already."

"It's not for you. You're making my eyes water just looking at you." Yoshi grinned.

Haroon aimed a playful cuff at Yoshi's head and returned to reading. He tried to quell the feelings of humiliation and resentment rising inside him. It wouldn't be fair to Yoshi to take those out on him.

Haroon wanted to concentrate on his book, but his attention kept wandering, and it was really hard to control his restlessness today. The Fujitas lived just a few blocks from the edge of J-District. By city standards, they were comfortable, but not wealthy. To Haroon's eyes, their tiny apartment was a palace. So clean compared to the dump he lived in. And more importantly, it was connected. He could access books, and art, and music, and courses, and even other people in faraway places. There were so many things to learn and he needed to know it all. It frustrated him no end that he always had such a hard time focusing.

He remembered the day he and Yoshi had met.

Haroon had been exploring since dawn. He'd been about ten and although it was dangerous to be out on his own, he had already worked out that if he was careful, his chances of catching a beating were usually less than they were at home, where there was no escape. And that afternoon, his curiosity was driving him harder than his fear. He had to find out if what the pasty-faced kid with the limp hair had said was true: J-District was a place, it had invisible borders, and things were very different on the other side of those imaginary lines.

Yoshi, meanwhile, had been out for a walk with his parents and somehow got separated. Running around in a blind panic, he'd gotten completely lost.

He was standing under a tree in an unkempt wooded park at the very edge of the District, crying, when Haroon had spotted him. Just a few minutes later, he spotted Dominic and Drea, the twins from the east side of his block who were also out, looking for trouble. It wasn't long before they heard the sniffling and started straight for Yoshi, grinning like a pair of sharks who had smelled blood.

Haroon's head pounded as he stood there, hidden by the shadows, watching as they got closer and closer, in an agony of indecision.

There was only one of him. And he didn't recognise the crying boy as anyone he might know or care about. But he'd seen what the twins had done to a stray cat once.

The memory still made him sick to his stomach.

He stopped thinking and charged.

The twins were so intent on their easy mark they didn't hear Haroon coming until it was too late. He threw his whole body against Drea, knocking her straight down, and then he spun and put everything he had into a hard right hook into Dominic's nose. He grabbed a stunned Yoshi by the arm and they ran as fast as they could out of the park.

It was several minutes before Yoshi had calmed down enough to think to check his wristband and work out where he was and lead them back to Yoshi's parents. They had been so grateful at Yoshi's quick and safe return they fed him the biggest meal he'd ever had in his life and told him he was forever welcome in their home.

Haroon remembered walking back into the District that night, blindly, a hurricane of conflicting emotions: wonder, dread, happiness, and furious resentment. Things *were* different elsewhere. And they were infinitely better.

He and Yoshi had become fast friends, and Yoshi's parents were helping him integrate into society and the thingweb, although not as quickly as he would have liked.

Mr Fujita startled him out of his memory by waving a plate under his nose. "*Anata wa kukkī o shitaidesu ka?*"

Yoshi rolled his eyes. "Dad, English, please. How many times do I have to remind you Haroon doesn't have a transbud yet?" he said, tapping his own, tucked into his right ear. With so many nationalities living in Toronto, the transbud was practically a necessity. And it was another thing Haroon had never been able to afford.

Haroon smile-flinched again. "It's okay, I can figure that much out." He took a cookie from the plate and nodded his thanks. Mr Fujita offered one to Yoshi, bent to kiss him on the forehead, and then left.

Haroon stopped in mid-bite. He'd seen Mr Fujita do that to his son a hundred times at least, but suddenly, for reasons he couldn't articulate, the image of that kiss seared Haroon. The room seemed stifling, and he could hardly breathe.

His hand flew to his bruised face, his father's handiwork. He was ashamed to find hot tears spilling down his cheek. Haroon wiped his eyes, tried to force his attention back to his book. He still couldn't focus.

If Yoshi noticed anything was amiss, he didn't say. Instead, he stretched, and then put down his own reader. "So," he said around a mouth full of cookie, "you wanna go do a holographic? I need a break."

Haroon didn't want to waste time downtown. He sure didn't want more distractions. Why was it so hard to keep his attention on one thing? "I don't know. I still have studying to do."

"*Uso tsuki*!" Yoshi scoffed at him and snatched his reader away. "Liar," he said again. "You were ready for the exams a week ago. You're reading more of your big ideas books, aren't you? Who is this one by? Seth… Bocci?"

"Bacchi," Haroon corrected him, before grabbing the reader back. "It's his second novel. It's all about how someone becomes so attached to their digital personal assistant that they start a physical relationship. And he seems to say that may not be as bad or as weird as you might believe, especially if you're not able to get along with other people. He's great."

Yoshi smirked. "Never heard of him. Some days I think you were born old. C'mon, let's flick. Maybe we can invite that cute girl you've been pining after? What's her name?"

Haroon sighed. Yoshi's continued friendship was important, and it wasn't such a bad idea to stop brooding for a while. "Fine, okay. She's called Saba, and she's more than just a cute girl, but no way are we inviting her. If I work up the nerve to ask her out, I don't need you telling her all of my embarrassing stories first time out."

Yoshi laughed. "Oh, I see, that's how it is. I'll tell her on your second date then. And your third because one date won't be enough for all of them."

"Yeah, and I'll tell that hunky football captain you've been crushing on him for months."

Yoshi fanned himself. "I almost wish you would. I'd die of embarrassment, but it'd be a happy death. He's gorgeous. But to do that, you'd have to get your nose out of the books for five minutes, so I know I'm safe."

Haroon chased him out the door.

SETH

Seth pushed back from his workstation and paced up and down. He had just wasted an hour drafting and then erasing a single paragraph repeatedly. Nothing was coming out right and the sounds from his old-fashioned keyboard, clicking and ticking, were irritating him. He considered switching to dictation mode, but he'd never gotten used to writing that way. Somehow, it felt less intimate, and too public, even if he was dictating alone, in his home.

"Tasha, give me some headlines. What's happening today?"

"At once. Scanning your favourite topics," Tasha replied. "Researchers in Spain have officially upgraded the status of the Pyrenean ibex from 'extinct' to 'extinct in the wild,' and noted they are now working towards creating a larger, genetically viable population for release."

Seth's brow furrowed. "I thought I'd read there was some debate about whether it was even possible to call it a Pyrenean ibex? That it wasn't, how would you say it, 'pure,' or whatever?"

"That is correct," Tasha replied. "Would you like to do a deep dive on this topic?"

"Not right now. I just need distractions to reset my head. Next headline."

"Four people were killed in unrest in parts of the African Union overnight, as representatives debate the pending trade agreement with the EU. Several world leaders, including China's new premier, Jiang Li Chun of the Chinese Social Democratic Party, have condemned the violence."

Seth shook his head. "Ugh, no, stick to positive stuff please."

Tasha paused as she worked to interpret that response. "Companions Inc. has announced orders for its latest creation, the North American House Hippo, have exceeded their projections."

That made Seth laugh. He'd seen the ads, and they were kind of cute. Miniature hippos, no bigger than a guinea pig, and engineered to have a nicer temperament than their dangerous wild cousins did. Plus no fur, so no shedding. He could imagine his great-grandmother complaining about them, though. She had never forgiven a neighbour for having two of those pot-bellied pigs that been popular once. "Another headline, please."

"The mayor of Calgary is asking the federal government to commission a study on droneway licensing. He says the number of new drones using the droneway between Edmonton and Calgary has tripled, but much of this is fly-through traffic en route to Lethbridge. He argues it is taxing the city's computational resources and the drones should be rerouted."

"Seems reasonable," Seth shrugged. "One more. Positive, remember."

"Climate Central is reporting that for the fourth year in a

row, carbon dioxide PPM has declined. Their spokesperson warns against decelerating current measures to fully realise a post-carbon economy, saying we are still nowhere near pre-industrial era levels yet. The switch away from coal for power generation and from fossil fuel for land-based transportation has been very effective, but much work remains to be done in the sea and air vehicles and in the agricultural sectors here and in developing countries. Reforestation is also still a concern."

Seth felt a little of the tension drain from his body. He still had nightmares about that awful documentary he'd seen as a child called *The Year of the Tornado*, which recounted how an unprecedented number of massive tornadoes had carved their way through the American Midwest in the late 2020s. The point of the video had been to show how that devastating summer, combined with a terrible disaster relief response and a history of low infrastructure investment, was the beginning of the end for the American union. But all Seth got from it was a permanent fear of storms, because Ontario was prone to vicious winds, too. Last year alone, Toronto had four ultra-severe thunderstorms, producing massive hail in some areas, destructive downdrafts, and flash flooding in much of the city. After each storm, there was always at least an hour's worth checking up on family members, as his large family was spread across the city. At least one of them would need help with property damage too. He wondered if he might look forward to fewer anxious summers.

With that hopeful thought, he sat down and got back to work. Soon, he could no longer hear the clicking of the keyboard, and the prose poured out. When he next checked his wristband, four hours had gone by. But it had been a solid afternoon. He'd added another two thousand words and written a tricky scene involving a major character reveal.

Feeling pumped, he had Tasha play music while he cleaned and danced his way around the apartment, belting out his favourite lines at the top of his lungs. He even threw open his windows for a bit to let some of the very cold, but fresh air in.

Then he made the mistake of sitting down again and logging in to check his book sales.

His sense of accomplishment vanished. Sales of his third novel had dipped noticeably, which meant his publisher's algorithms had stopped showing the book to potential buyers. It was plummeting into obscurity, destined to languish near the bottom of the rankings in every possible category. Just like the first two.

A horrible thought struck him… He ran a report comparing sales of his three books over time. The charts were worse than he feared. The sales decline was happening sooner after launch with each new book, and that could only mean the algorithms were giving up on him faster, based on historical results.

He got up and kicked his chair. It had taken him years to get a publisher. Editor after editor had refused to even process his manuscript, claiming they didn't think their software would rate it well based on his synopsis. "As I'm sure you appreciate, we have a limited budget for text analysis, and we receive tens of thousands of titles per month" was how the rejection message usually read.

So when Seth made it past both an editor and the software and signed a contract, he thought he was set. And then they'd simply thrown his first book into the marketplace to sink or swim. That's when he'd learnt that if a book didn't sell at a certain rate within the first week — one week! — it wasn't allocated any marketing budget. And if it didn't get any funding, it sputtered through the new titles lists and then the recent release lists, becoming less and less visible until it couldn't be found unless a reader searched for it deliberately.

He fumed for a while longer, and then paced again, his jaw setting firm. He had always resisted going independent, because it seemed like a lot to learn and deal with: formatting for fabbers, formatting for the virtual editions, distribution, learning to use blockchain to sell and protect the book from piracy, learning marketing algorithms. Time was already hard enough to come by, what with the near-constant demands of grandparents, parents, siblings, cousins, nieces, and nephews, who all seemed to think he was both unemployed and in need of a spouse. Seth had hoped to stay focused on writing with what little time he had to himself.

But not selling books was even worse.

Seth caught sight of his ancient workstation and laughed at himself. What was it his sister had once called him when they'd talked about his dating life and refusal to use a digital matchmaker? A fuddle-duddle? No, a fuddy-duddy. She'd called him a hypocrite too, as he was happy enough to use technology to obsess about his health. Dammit. Why was she always right about this kind of thing? Here he was trying to craft a book about someone casting aside the modern tools for creating and disseminating art, and yet Seth hadn't explored what was available to him in real life to truly understand what he was rejecting. No wonder he was struggling to write something authentic.

Well then. He squared his shoulders. If the choice was sink or swim… it was time he learned how to swim.

MEIKE

The bar countertop in the club was huge, extending perhaps ten metres to either side of where Meike sat. She watched the barbots, multi-armed automatons at their stations, swiftly pouring shots, muddling ingredients, shaking and stirring drinks for the endless lines of thirsty partiers. They were mesmerising, especially since they were weirdly synced with the music.

The bass was intense. The dance pit was an enormous circle with a floor dusted with multi-coloured sand that vibrated into cymatic patterns in time with the beat. Whenever it looked like the action was slowing down, virtual dancers would materialise, showing off new moves and working the crowd into a fever pitch again.

The club was the place for the wild and the exhibitionist. While most residents of Toronto were conservative about their upgrades, choosing invisible enhancements to things like mental performance, strength, or endurance, the people here wore them loud and proud. Most of it wasn't legal, or at least, not officially reviewed and sanctioned by the Ministry of Health. If you liked what you saw, you'd have to ask around to find out which cosmetic surgeon or tattoo artist was moonlighting as a modder.

Meike was drinking something that was purple and foaming violently down the side of her glass. She couldn't remember if it was her fifth or sixth drink.

Someone slid into the chair beside her. She had a shaved head, pointed ears, and tiger-striped eyes with dilated pupils. When she smiled, her teeth all had points.

"Heya, new girl," the stranger said. "Don't think I've viewed you before." She made of a show of looking Meike up and down. "You're a Plain, ya? Nothing hidden under those prettay, prettay clothes? I can fix that, sure ya."

"No, no thanks," Meike sipped her drink. She knew she couldn't afford anything this woman had on offer.

"Oh, come on now, let me at least do something for your eyes. A lot of the girls like you love the dragon eyes."

Meike looked around. She hadn't realised before but there were plenty of dragon eyes in the club, some bright red, others violet. She shrugged. "Not into mods," she lied.

The woman's eyes narrowed as she considered Meike. "No mods, heya. Okay fine. You not a biopunk or grinder, I don't see no implants, and you def not a cyborg, so. Augs? You an aug? What you want I can score for you? I got all the access."

Meike took a longer pull of her drink. She had never seen the point in augmentations. Her job didn't require things like strength or speed, and thanks to her androgynous features, which men always seemed to find fascinating, she'd had no trouble getting sex partners. Bigger body parts had no appeal.

After a moment of silent thought, the woman said. "Bet you like some Feels?"

Meike suddenly focused on her. "Maybe. What do you have?"

The woman laughed knowingly. "Ya, I thought was so. It's all about the experiences these days! How about some electrics? You ever try that? Just tell me for certain sure you don't got the implants so I don't mess them up good." When Meike still hesitated, she added, "Is very cheap, ya? So basic."

Meike could feel the bass vibrating in her chest and noticed the taste of her drink for the first time that evening. "Sure," she said.

The woman made a gesture signalling for money. Already itching with excitement, Meike flicked a few screens on her wristband to pay the bill, and then took it off, laying it on the table so it wouldn't be damaged.

Transaction completed, the woman pulled a small box with a simple silver dial out of one pocket. Out of another, she took out a packet of electrodes. She paused. "You want here, or I got a room in back?"

Meike looked around. At the next table, a short, black male with phosphorescent tattoos and super-extended fingers on each hand was chatting up another man — or was it a woman? — who had their whole face covered with electric-blue scales. No one here would care.

Meike motioned for the woman to proceed. Moving with astonishing speed and grace, the woman's hands snaked under Meike's clothes here, and there, placing the electrodes on some of the softest, most tender parts of her body. She then set the box in front of Meike and turned the dial to the first position.

Meike stiffened as the current hit her. She sat there for a moment, tingling, her breath quickening, and her eyes wide. Then, without warning, she gulped down her drink, grabbed the box, and cranked the dial to the top position. She stood straight up, shrieked once, felt her lips pull back into a rictus grin, and then fell down hard, convulsing, jerking and twisting herself into a ball.

The safety kicked in and the device shut down. Her whole body shook and her chest was heaving as though she'd run a marathon. She tasted blood where she'd bitten her tongue, and her head throbbed. For a few fleeting moments, she felt everything. And then it all faded.

Meike looked up at the woman, who had a huge smile on her face, her pointed teeth gleaming in the flashing club lights. Some people nearby had seen Meike go down and were lining up for a turn of their own.

She needed more money. More of this. More *alive*.

KEL

Kel sat alone in the cafeteria, moving her spoon idly through her soup, watching it swirl and eddy as the steam curled into the air. She was hungry enough, but the soup was too salty, and that made her think of the paper she'd just read about sodium channels in axonal conduction and…

She threw down the spoon and tried sipping her coffee instead, hoping maybe the caffeine would clear the cobwebs. It was so hard to concentrate. Sure, there was the paper to finish, and she no closer to finding out what had happened to her dead macaques or her data logs, but she still had all the others, so she should be busy. The data from them — which she was now backing up in six different places, both offsite and onsite — was ticking in quite nicely. There was still so much to be done.

Kel thought of her grandmother, Madeleine. Smart as a whip and acidly funny, Gramma Maddy had started her career as a teacher but quickly became impatient with the curriculum and the system, as she felt it wasn't doing an adequate job of preparing her students for the future. She'd thrown everything she had into her work, rising through the ranks to become a principal, and then a superintendent, and building support and relationships in the teachers' union. At age fifty, she ran for public office, and won her seat in the provincial legislature handily — thanks in no small part to the friends she had in the union. Within three years, she had a cabinet post as Minister of Education, where she set about overhauling the Ontario curriculum, and won a seat again.

She would have made premier, Kel thought glumly. Or maybe even gone into federal politics and been prime minister. Who knows what she might have accomplished then? But at age sixty, disaster had struck: the first signs of early onset Alzheimer's. Kel had been thirteen. Madeleine faced it bravely and retired from public life. The disease had been swift and ruthless: Madeleine had not seen her sixty-fifth birthday.

Kel rubbed her eyes. The memories of that savage decline still haunted her. It was her greatest wish to prevent that kind of suffering in all people, but especially those who had dedicated their lives to making things better. Kel hated to see potential like her grandmother's cut short. She also secretly worried about her own future. The causes of the disease were still not well understood.

A shadow darkened her table. She looked up and tried not to groan. It was Bao-Yu, and worse, Padraig was with her. At a little past ninety, Padraig was due for retirement any day. She wished he would. He drove Kel nuts, as he seemed to spend all his time here wandering around and chatting with people, interrupting whatever they were working on. Bao-Yu was always asking her strange personal questions, but at least she did it on breaks.

"Such a frowny face," Padraig said, sitting down across

from her without waiting for an invitation. "You should smile more, you know."

"So I've been told," Kel said, resisting the urge to plaster on a fake smile to appease him. "But it's been a rough week."

"Well actually," Padraig said, "the sun has been shining and spring isn't far off. So really, there's nothing to complain about." Bao-Yu chuckled, as she too sat down.

Kel bit back a sharp reply, knowing it would only bring out an hour's worth of unsolicited advice. She moved to stand. "Yes, well. I should probably enjoy some of the sunshine. If you'll exc—"

"I heard you had a spot of trouble with your baboons?"

She gripped the handle of her coffee cup a little tighter. She knew if she left now, Padraig would take offence and complain to Robert, Bao-Yu would back him, and also make the next staff meeting exceptionally difficult. Staff meetings already went on forever. But she wanted to try to get some work done as the caffeine worked its magic.

"Cute little apes, I always thought." Padraig smiled up at her though his eyes didn't crinkle.

"Macaques," she said as she relaxed back into the seat again. "They're macaques. And yes, two of them died, out of the blue."

"That's terrible!" said Bao-Yu. "Did you lose much data?"

"A lot of potential data," Kel replied. "It won't cripple the project but it is demoralising."

"They're not as interesting as, say, *Shinisaurus crocodilurus*, but cute," said Padraig, carrying on as though she and Bao-Yu hadn't been talking. "You do understand what that is, don't you?"

"Yes, the Chinese crocodile lizard. You've mentioned a few times before," Kel said.

He folded his arms across his chest. "Remind me what your little project is about again?"

"My long-term *study*," Kel said, trying not to overdo the emphasis on the last word, "is looking at monkey and ape resistance to tauopathy and AD-related neurodegeneration. Or in other words, they get senile plaques, same as we do, but don't seem to have the same issues we do. I'm studying brain development and degeneration in real time."

"Ah yes, your implants," Padraig said, raising a finger. "That will be the problem." Bao-Yu looked at him quizzically.

"I'm sorry?"

"Implants. Never a good idea. That's why your monkeys died."

"But all of them have im—"

"All this modding and augging and such like. Messes with the head. There's a reason our brains are encased in skulls. It's as much to keep stuff out as to keep other things in. Bypassing that and the blood-brain barrier? Flooding the brain with nanoparticle transmitter agents? Recipe for disaster. Your monkeys probably went nuts and killed each other. We gotta stop messing with this sort of thing."

"You realise," Bao-Yu leaned over to Padraig to nudge him conspiratorially. "Kel doesn't have any implants. She told me once. She's that smart naturally! Isn't she lucky?"

Padraig snorted. "So you won't have any yourself, but you inflict them on your animals?"

Kel blinked. "What? Super-high resolution tomography is safe and has been for years. And I thought you were a scientist? How can you not want to know what's going on inside and out?"

Padraig rose stiffly. "There's no call to be that way about it. And anyway, I don't see how I could do anything like that, what with your never-ending torrents of data making everything in this facility move like molasses in January. I don't understand why you need to record every single thing each of those little ape brains does every day." With that, he huffed out of the room.

Kel sat there feeling stunned, not knowing what to make of the conversation. Did he really not approve of implants, or was that just an excuse to complain about Kel's resource use? And that was the second time she'd had someone mention that. Did all the staff feel that way?

Bao-Yu reached over and patted her arm. "Oh, don't mind him. You know how he gets if he hasn't had his afternoon tea. But heaven forbid you should remind him to drink some, or he's worse."

They laughed and parted company, with Kel headed to her desk. The whole thing left her unsettled though, and she kept sneaking looks at all of her co-workers. Just how strongly did any of them feel about her work?

MAURA

Coming back into her office, Maura was disturbed to find an alert flashing at her desk. Someone had been in her office while she was out.

She sat down and logged into her computer to access security footage. To her astonishment, it had been Pauline. When had she slipped out of the meeting to come back here? Why?

In the video, Pauline entered Maura's office, looked at her wristband — noting the time, maybe? — and then glanced around the room, as though debating something. Then she walked across the office to the wall opposite Maura's desk to take a close look at the Van Gogh print there. She consulted her wristband again, this time asking it about the print, querying as to whether it was an original or not, and who it was by. She did the same thing with the other two art prints, studying them for a few minutes, and then she walked over to Maura's desk. She examined the sculpture on Maura's desk closely.

Maura leaned forward at her desk to watch the image of Pauline try bringing up Maura's computer. It wouldn't respond. Maura nodded, satisfied. The computer wasn't even visible on the office network, making it impossible to access from elsewhere, and Maura had it set up to appear only for her, even in the safety of her own office. She watched Pauline drumming her fingers on the desk.

Pauline went to the fabber next. While Maura's tea preferences were well known within the company, she had never asked her assistants to get anything else. Pauline accessed the fabber settings, and — Maura zoomed in — checked out the order history. Interesting. Maura ate breakfast and lunch here most days, but never dinner. Pauline would have found frequent prints of arepa and cachapa, among other things, both of which were Venezuelan. Maura rubbed her chin. She made a mental note to check outbound queries to Canadian immigration services to see if Pauline was looking up her immigration records.

"You shouldn't be in here," said a voice from the door in the video. The suddenness of it made Maura jump even though she wasn't the one snooping, and she laughed at herself.

Pauline turned to see a security guard glaring at her. "Oh, hello," she said casually. "Yes, I'm her executive assistant. I don't believe we've met. I had a problem with this unit earlier, and I was just taking a moment to see whether it was my error or whether I need to get it serviced." Smooth, Maura thought. Cool under pressure, too.

The guard glared even harder. "Yes, I know who you are. And Ms Torres has no exceptions to the rule about visitors to her office in her absence. None. Get out."

"Oh dear," said Pauline, taken aback by the guard's

rudeness. "I'll have to talk to her about that. In the meantime, though, we don't want you getting into trouble, do we?" She smiled as she walked to the door. The guard's stance did not soften one bit. Maura made another note to herself about a surprise gift for the guard.

She watched Pauline leave the office. Maura quickly reviewed the camera views in the hallways and saw Pauline's retreating form head back to her own office, close the door and lean against it, puffing out her cheeks with what looked like a relieved breath. Then she straightened her clothes, exited, and apparently returned to the meeting they'd all just attended.

Maura checked the footage of the brainstorming meeting. Pauline had left when they were all engaged in their discussion about new titles for the games division; her absence was no longer than a trip to the bathroom would have been, and Maura sensed that even without the interruption, Pauline wouldn't have been very much longer at it. The question was, why hadn't she thought about video surveillance? Pauline seemed too smart to get caught out by something that obvious. And why investigate artwork, of all things?

It was a puzzle. And Maura wasn't sure whether she liked it or not.

RAY

Shivering. Teeth chattering. A hand like a lead weight on his shoulder, shaking him.

A woman leaned over him.

"Wake up. Come on. Snap to."

A slap on the face. He groaned. His ears rang.

"Officer! What on earth do you think you're doing? There's no need to—" a voice said

"Get lost. Go nurse someone." The woman.

"Mwita, find the ward superintendent. And security. Quickly!"

Another slap. He opened his eyes again. The female officer glared down at him.

"Who are you?" she asked.

The room spun around him.

"What were you doing downtown?"

"This patient is in no condition for interrogation!"

"Did you set off the bomb?"

"You said you were just going to look at the patient!" said the other voice. "Do you even have authorisation to do that much? This can't be standard procedure! Why are you being so rough? What's your sergeant's name?"

Somewhere off to his left, there was a machine, beeping

at him again. The beeps were coming faster.

"I don't like it when things blow up on my beat. Who are you? We haven't had this kind of nonsense here for years, and I'm not about to let it happen again. Who are you?" the officer demanded.

In the distance, there was shouting.

The woman leaned closer. Her uniform was bright blue. She held a tablet up to his face. It displayed an image of a body.

"Who is this?" she shouted.

Mick… Most of his hair and clothes had burned away. An arm had sheared off at the shoulder. He could see ribs. Part of the pelvis, shocking white against the red and black mass.

The room stopped spinning, and an ice-cold ball of sick formed in his stomach. He gagged, and the officer stepped back in a hurry.

He gagged again, and the gag turned into a retch and the retch turned into a gasping dry heave. He tried to turn over, to curl into a ball, but he couldn't. The sheet over his body bloomed red in several places.

He couldn't stop heaving and then he couldn't start breathing.

More people. Shouting. A scuffle. The beeps blended into one long whine.

Blackness.

MEIKE

It was just after four in the afternoon when the low sun slanted across Meike's face, making her groan. Her head pounded and her mouth tasted like warm copper and electric tingles. She flung an arm over her eyes and lay there, cursing whoever had sold her whatever it was she'd tried last night.

Eventually, she sat up. Her room was the same beige colour it had been when she had moved in. There was a cheap dresser, a cheaper end table, and her bed; her walls were devoid of pictures and decorations. Her bathroom and kitchen were also bare and plain.

Two floors below her apartment in the basement, the deejay in Club Rax fired up the night's first set of binaural beats. The building thrummed in resonance.

Meike stumble-slid out of bed to the window to look down. The stream of pods along Church had slowed to a trickle as people jaywalked from one edge of the flow to another, and the sidewalks were filling up.

On the far side of the flow was a long, low row of tidy modder shops with enormous windows, each painted a different colour of the rainbow. In Jackson's, a Baxter unit was airbrushing a sunset onto a man's bare chest. A crowd gathered outside the window at Digitz to watch a teenager get magnets implanted into her fingertips. At Black Eagle, the night's featured rentbot was a blond male; threesomes were on discount.

Meike felt a wave of something that could have been nausea, or maybe hunger. She pulled on some clothes and slipped out of her apartment, taking the lift rather than trusting herself on the stairs. Cold, damp air slithered into her clothing the moment she stepped outside, the kind that always made her ache.

Kaiten Noodle beckoned, mostly because there wasn't a line yet. The interior was garish: all bright red and lemon yellow and black. The room was huge and tightly packed; a long, continuous conveyor belt-track snaked and looped its way through the restaurant, beginning and ending in the kitchen. The tables squeezed up against it.

The welcome detector at the door identified her as a party of one and told her to go to table seven. As she sat down, a section of the table angled up, offering her a touch screen filled with brightly illustrated food choices. She ordered champon and a triple Suntory whiskey, then waved her wristband across the scanner to pay. As she waited for her meal, she alternated between rubbing her temples with her fingers and dragging the knuckles of her thumbs across her eyebrows to ease her headache.

Her meal made its lonely way out of the kitchen on the conveyor belt. She picked it up as it trundled close, inhaled the steam.

"That looks good, which one is it?" said a voice behind her.

She turned around. A young man, perhaps Chinese or Malaysian — or possibly both — was smiling at her. He wore a few days' growth of facial hair and had a sweet face.

Meike went back to her food. "Number six special."

"May I join you?"

She shrugged. After a moment's hesitation, he sat across from her. He ordered the same as she had, and then sat back, looking at her closely. "Do you remember me?" he asked. "From last night?"

Meike stopped eating long enough to inspect him. "Could be," she said. "Maybe. I was pretty hacked, I think."

The man nodded. "You were. My name is Fa. We met at the club. We had a good time." His order reached the table, so he stopped to grab it. "You, uh, well, you were saying some pretty brass things."

Meike slurped up noodles. "Was I?" she said. "Like what?"

"'The lie is a condition of life' was one. And something like 'We delude ourselves. Our lives are of no significance.' You said a lot more, but those are the two lines that stuck. It all sounded crazy deep."

She raised her glass and threw back a gulp of whiskey, closing her eyes to savour how it burned in her throat. When she opened them again, Fa was staring at her with frank fascination. "Not anything I wrote," she said simply. "Just stuff I've read. Sartre. Nietzsche."

"Philosophers, right?" Fa asked. She nodded. "You've read a lot of it? Philosophy, I mean?"

"Pretty much all of it. Studied most of the world religions, too."

"Wow. Even Voodoo?" He grinned.

"Yeah. And Taoism. And a bunch of others."

Fa scooped up some of his meal. "Which ones did you like?" he asked between bites.

"Who knows? They all sound the same to me," she replied.

He picked up his drink, something bubbly, and stirred it a few times with a stainless steel straw. "I've been reading this guy named Campbell. He's talking about all the common elements in religions and the hero's journey. I really want an epic journey of my own. I'm trying to get tickets to this year's Burning Man. So hard to find since they clamped down." He looked at her. "You read this stuff for school or work?"

"Nah. I'm just a lab tech."

"Why did you become a lab technician?"

"It pays enough to buy drugs," she said flatly.

Fa choked on his drink and laughed. "Really? Are you serious? You're something else." She shrugged again, nodded, and ate some more dinner.

He talked for a long time, about his parents, about how he wanted to learn about this Sartre guy, and how he'd been thizzing last night on Ecstasy. He described a book about perception by Huxley, which he said made him want to try peyote now. She let the words wash over her and ordered another triple whiskey.

When she finished, Meike leaned back and considered him. In her whiskey-fuelled haze, she wondered how you could read Campbell and get nothing but a yen for festival tickets out of it. But he was quite lean, a little on the short side, and well muscled. His hair was a deep black and very thick; he'd let it get wavy at the back. He smelled good. And it had been a while. So when he finally got around to saying he wanted to get to know her better, she said yes. They went to his place.

KEL

There.

Kel slumped back, at once thrilled to have found something yet dismayed it was there to find.

The sun was rising outside. She was alone in the lab again as she hadn't gone home last night. On the screen was the implant log for Max, showing thousands upon thousands of data points, all transmitted by the implant to the lab computers.

All inputs, except for one line. An output. Somehow, her implant had been used to send something from the transmitter to the brain and it changed the macaque's behaviour. Pure noise, it looked like, and she guessed it would have been just enough to cause Max to freeze in confusion. There were several of these, scattered at random intervals throughout the log, but this was the last one, and it happened not long before the implant ceased transmitting.

She reviewed the time stamp for the signals and checked the corresponding point on the video logs. Max wasn't always in view of the camera, but whenever he was, he paused at exactly the same moment the signal hit him. Each time, for a few seconds afterwards, he looked around vacantly, like someone who had come into a room but forgotten what they were there to do.

Max was a swinger, spending nearly all day flinging himself from tree to tree. He'd plunged to his death when one of those outputs had struck.

Now knowing what she was looking for, she checked Dalton's log and found identical signals. Indeed, there were more in Dalton's records, but then again, he spent more time on the ground. It would have taken longer to kill him.

Kel paused, stunned by her own thoughts. At the moment, all she had was evidence of something happening with the implants. What made her think of the word *killed* rather than died?

She took a swig from the mug on her desk, and made a face at the cold coffee. It was the same feeling that had prompted her to look at the implant logs again, she mused. Padraig's antipathy. Two macaques dying suddenly. The missing log for Pika. She recalled a conversation with one of her undergraduate professors. She'd just shown her the results of tests for a new radiographic contrast agent she had proposed, which would be less toxic than the industry standard, and cheaper, too.

"This is good." Professor Ramesh had nodded. "The methodology looks solid. Go ahead, write this up, and show me the draft. We'll see about getting it submitted. I wouldn't make a big deal out of this in class though."

"Why not?" Kel asked.

Ramesh cocked an eyebrow at her. "This isn't typical of undergrad work in general, much less someone of your age. At least two-thirds of the class are only there because it's a requirement. If any of the rest of them are looking at grad school and becoming researchers, they will resent you for beating them to this kind of original research. As it is, surely you've noticed they aren't exactly thrilled to have you in class? You always know the answers." Kel had puzzled over that for a while. With all the problems in the world, why wouldn't they be excited over something that would help make things better, even if only in a small way?

As for how they treated her, Kel hadn't noticed. Her mother had chastised her more than once for being unable to provide information about her classmates, what they looked like or what they did in their time out of class. But Kel had never really seen the point in paying attention to such details. In her view, lecture halls were temporary gatherings of people she was not likely to see again once the course was over at the end of term. Why waste time with gossip? But now, she was second-guessing herself... Had they felt like Padraig did?

She shook her head, berating herself for being foolish. Everyone knew Padraig was prickly, even more so now with his retirement looming. This was a lack of sleep and stress making her overthink things.

Robert blammed through the office door. His eyes narrowed as he spotted her at her desk. "I took a breakfast meeting with a friend of mine in the Ministry of Finance," he said, biting off every word. "And rumour has it big cuts are coming unless each project can show they're working towards a commercial application or a significant public-private partnership."

"What? Wasn't the last federal budget running a surplus? What do they need to make cuts for?"

"I couldn't begin to guess." Robert ran a hand across his face, rubbing his eyes. "It will be this government's first budget. My friend figures the thinking is to take credit for the surplus of the previous government by making it even bigger for the first few years and then spending all the savings on a big and popular initiative right before their term is up. Or something. Who knows?" He shoved his hands in his pocket. "At least we've had a tipoff. But what it means is the last three weeks I spent on our budget were a complete waste of time, and we've got less than a month to come up with something plausible. Staff meeting in an hour."

RAY

It was nice here, in the darkness. Warm, and the pain was only a background noise, like rushing water.

He'd been thirteen when his brother Cedric was born.

Things had been okay for a while, just then.

Because of Peter. A good man.

Peter, who had seen… something in Mom and decided to try to save it. Ray had never worked out what to call Peter. Uncle didn't seem right, and neither did stepdad. He regretted it now, not giving him a proper relationship name.

Kind eyes, a firm jaw and a steady job. Hope. Dreams. Peter. Hints that things were different in other places. And you could leave here and go to those places.

Laughing! Peter laughed. A lot.

Eating every day!

Mom did not laugh. But for a while, Ray didn't want to hide from her all the time.

A day in April. Rain outside. Ray watching the birth from across the hall in his bedroom. His mum, screaming in agony, begging Peter to make it stop. He wondered if this was why she hated Ray. Why hadn't they gone to a hospital? He had never thought about it until now. Perhaps because his mum was sure the spooks would get her. She was always talking about them, although Ray didn't know what those were.

Cedric, they called him. A tiny, squalling, almost blue thing.

Peter and tears of joy.

Ray holding Cedric. Terror at the thought of holding a… a… someone in his arms. Once, when no one was looking, bringing Cedric extra close and feeling his heart beating. Who knew babies were so soft?

That day.

Coming home, after a day in the May sunshine, feeling like he'd never felt. Like maybe, just maybe, good things could happen.

Peter, on the floor in the centre of the apartment, moaning, crooning, tears and snot streaming down his face, cuddling Cedric, stiff and still in his arms.

Ray knew, almost immediately.

His mother was using again, and it had been Cedric who had absorbed all her rage.

He should have stayed home.

He was bigger than Cedric.

MAURA

Maura's spacious personal pod swept silently down the Bridle Path and then slowed to turn into the entrance of her estate. The black wrought-iron gates slid open, allowing the pod to follow the curve of the driveway to the front door where it pulled to a stop. It waited a few moments before making a polite ping to alert Maura they had arrived.

She looked up from her work, surprised as always by how short the trip seemed. She bent over to pick up her bag, her synth leather seat creaking. Maura slid a hand along the burnished wood accents of the handle, found the button, and the pod door purred open. She got out and headed for the house while the pod slipped around back to park itself in the garage. The front door opened at her approach, and the lights in the bright foyer came on for her. She let out a little sigh of contentment, as the sight of her home coming alive to greet her never failed to make her day.

She'd fallen in love with the place as soon as she had seen the listing. It was a house of geometry, all rectangles, and straight lines, and symmetry. The outside was faux-Edwardian, while the interior was early twenty-first century, old enough now to be retro chic again. The palette was soothing creams and warm, dark browns. It gave her a sense of solidity, and order, and spoke of wealth without being as ridiculously ostentatious as some other homes on the Path. And most important, it was so very different from the chaos of her own upbringing.

She set her bag down on the lone, oval table in the foyer and walked into the main hallway, her hard shoes clicking on the marble tile. She took the staircase, a minimalist curving work of art all by itself, up to the second floor. Her DPA had already embodied itself in its mobile form and was waiting for her in the master bathroom, having laid out a fresh towel and her housecoat. The humanoid unit was dressed casually, in pale slacks and a long-sleeved blue shirt open at the neck. The bathtub was filling itself, the water heated to the temperature she preferred.

"Jarvis, what do we have in the way of Malbec tonight?"

Jarvis' head tilted slightly, indicating he was accessing the inventory. "We still have a few bottles left from the case of Catena Zapata. Should I get one?"

"That would be perfect. Decant a bottle on the kitchen table, please. And pick a paired selection from the fabber to go with it. I'll be in here for about twenty minutes."

Jarvis nodded and walked away, his servos whirring softly. She undressed and stepped into the bath, which had stopped filling. When she settled back, the massage jets came on. She willed her muscles to loosen up, letting the water swirl around her.

She considered her options. The CEO of Imprint Tech had rebuffed her first thing this morning, refusing to even take a call from anyone at EduTain. His reaction puzzled her. She knew from their agents inside the company that second-quarter results would be about four times worse than market analysts had predicted, which meant costs were spiralling out of control there. That would not go down well with investors and would result in a big sell-off. What she couldn't figure out was why costs were so high. None of their contacts reported a hiring spree, any major investment in hardware, or content development projects. And yet, several of their bestselling medical training simulations badly needed updates.

What happening there? Where was the money being spent? She wondered if the CEO was gouging the company, taking the profits all for himself. The business celebrity chatter websites didn't have any headlines about him engaging in any lavish spending personally. There were no lifestyle-section photos of him carousing on a new private island, for example, but then again, they weren't always reliable sources of gossip. Perhaps he was funnelling money offshore.

Maura took a dim view of that sort of behaviour. It was one thing to live well, if you'd earned it, but quite another to gut the balance sheet and endanger the livelihoods of the people who trusted you and worked for you. While some executives at her level might own several houses, Maura kept just one, and if pressed, she'd happily admit she got it for pennies on the dollar at auction, the same way she'd acquired most of her possessions. There were so many interesting things to spend money on: sponsoring scientific studies, helping to fund the planned Mars expedition, or even reforesting the southwest portion of the province.

Maura reached up to the shelf along the tub insert and found her favourite bottle of lavender oil; the only scent that didn't make her sneeze. She put a drop in the hot water and inhaled. That left the president of Xperience to woo, at least for now. Cheryl Bhattacharya was a darling of the virtual reality gaming industry, having built a solid franchise in the fantasy role-playing market practically from scratch. She had done it through meticulous research — legend had it she'd personally attended every major fantasy con on the planet for five years in a row — and shrewd poaching, luring staff away from old stalwarts like Ubisoft and Blizzard. She had worked out exactly what people wanted: smart, immersive, challenging worlds with just the right amount of fair competition, and got the best talent to produce it.

Maura smiled and sank lower into the water. She felt she would get along well with Cheryl, as her own ascent had been similar. When Maura had first inherited EduTain, it was a nascent software company in the augmented-reality sector. While other companies in that space killed themselves trying to produce the next breakout game or intrusive advertising platforms, she had focused on the tourism industry, producing highly detailed and fun AR overlays for major tourist attractions. Her software succeeded where others failed because she'd done her research and figured out different tourists wanted different things. Whereas one tourist might want to grok out on the construction details of the Great Wall, another would love to know how it fit into the grand sweep of geopolitical history. EduTain's code, delivered through stylish wearables, customised everyone's experience.

With revenue from that pouring in, Maura could then take her time, acquiring other equally unsexy but robust businesses in AR and then VR. The question was what carrot to dangle for Cheryl to get her to merge with EduTain? A leader like that would want more than money. She would need a challenge.

She finished her bath, slipped into her housecoat, and padded through the big, silent house down to the kitchen. Jarvis had laid out her food and wine and set up a virtual screen with her messages. The first one made her smile: Councillor Brown's polling numbers were already down five percent.

As she tucked into her meal, only one thing prevented her from relaxing completely. There was another question bothering her this evening: Was Pauline a spy now embedded in EduTain?

KEL

Her messenger software dinged again. Kel groaned. There could only be one person who would call her this early in the morning.

"Answer," she mumbled, sitting up in bed. The messenger beeped brightly in response. "Hello, Mother."

"Hello, dear. Where's the video?"

"Mother, it's 5:30 a.m. I really don—"

"Oh, don't be silly sweetie, your dad and I have seen you this way before."

Kel sighed and did her best to wake up and smooth down her hair. "Video," she told her messenger.

An image of her mother and father appeared. She tried not to think about how baggy-eyed and rough her image would look at their end.

"There you are," her mother said. They were both in their kitchen, standing stiffly in front of the one camera they owned, dressed as usual in clothes that were at least six fashion cycles behind. Canadian Gothic, Kel thought sourly.

"Hello, Kelleen," her father beamed. "How's life in the city?"

"Father, please, I haven't been Kelleen since I was four."

"Just checking to see whether you were awake. So..."

Don't say it. Just once, don't say it, Kel thought.

"…why are you still in bed? Half the day is gone!"

Kel sighed.

"I had a late night, guys," she said, and instantly regretted it.

"Oh!" Her mother brightened. "A nice young man?"

"Aaaarrgggh, no!" Kel threw the covers off and staggered towards the kitchen. She had more important things to do than get involved in the drama of a relationship right now. The image of her parents reconstructed itself over her counter. She fumbled for the coffee. "I was working," she said over her shoulder to the display.

"They seem to work you hard at this job," her mother said. "Any prospects? Is your assistant cute? What about your boss?"

Only the scent of fresh coffee brewing prevented her from shuddering at the thought of going out with Robert. "Most emphatically not," she said, and, desperate to change the subject, she continued, "So, what's new with you?"

While her father launched into the latest gossip from his game club, she tried to suppress the growing anger she always felt when her parents called. *You are such throwbacks!* she wanted to shout at them.

Her father was a fabber tech, in charge of inspecting, calibrating, and maintaining a series of the big, central fab stations throughout Muskoka that printed pods and housing units. It was all he'd ever aspired to. Her mother had finished her mandatory education, found her father, and promptly gotten pregnant. She had never even tried to enter the workforce. Kel suspected her mother would have been an Analogue, but for the fact she liked the creature comforts technology provided too much. She'd shown no desire to participate in modern society.

They hadn't known what to do with Kel. Precocious from the beginning, she had grown bored with her toys quickly and pulled things apart to see how they worked. They had done their best, supplying her with unlimited access to the thingweb far earlier than usual, taking her on trips, allowing her to sign up for all the extracurricular activities offered by the local school board, but this just seemed to fuel her discontent and make her want to do more.

As her father moved on to relaying the latest news from his job, Kel recalled one terrible day, not long after she'd turned ten. She had discovered a documentary about the poverty and extreme pollution in Louisiana and confronted her parents as to why they hadn't done something about it. They hadn't understood why someone would be so upset over an event they couldn't control, and which wasn't even local. She didn't understand why they wouldn't try to end suffering, no matter where it was. They'd had a huge fight, and Kel hadn't spoken to them for days afterwards. It had been the first of many blow-ups.

It had been a relief to everyone when Kel eventually left home and moved to Toronto.

And yet, they keep trying to get me to come back, she thought, as she took a deep pull of coffee. For the first few years, the only thing that had brought her to Muskoka was her grandmother. It had been all Kel could do to attend her funeral.

"—and you're not listening."

"What? Oh, sorry, Mother. It's just been a very, very rough week."

"Oh dear. Tell me how."

Kel tried to resist; she wanted to brush her off and yet… this was her mother. It was so hard to ignore the immediate softness that had entered her mother's voice, the virtual hug Kel was being offered. Before long, she found herself pouring out the last month's woes, her parents looking concerned, nodding in sympathy at all the right places.

"Did you write a protocol to prevent further issues with the other implants?" her mother asked. "Perhaps something in COGOL to make it hard to detect, much less work around?"

The warm feelings that had been rising in her quickly vanished. This, *this* is what Kel could not fathom. One minute Mother would be prattling on about her back garden, the next she'd let slip something that proved there was so much more under the surface. Why did she suppress it? Why didn't she use it? Such a waste of potential. Kel put down her coffee cup slowly and deliberately, willing herself not to shout.

"Yes," she replied. "Or at least, that's what I was working on last night. It's just all very frustrating as those were my two oldest macaques and they were just about to produce some senescent data. I should have thought to safeguard against something like this."

"My girl," her father said fondly, "the manufacturers of these kinds of things are supposed to have all sorts of quality controls."

"Yes, you can't be expected to think of everything," her mother agreed. "Stop being so hard on yourself. It's nearly impossible for one person to make a difference in today's world."

Kel bit back many replies and politely signed off. She had enough on her plate without feeling guilty about another fight with her parents.

HAROON

Yoshi threw himself into the chair and aimed a kick at the cafeteria table. Haroon lost his grip on the reader he'd been holding — another Bacchi novel — and it skittered to the floor.

Haroon gave him a *what the hell?* look, then bent to pick up his reader. Yoshi sat there sullenly for several minutes before answering.

"Sorry. I'm just completely done with my father," he announced.

"What's up?"

Yoshi rolled his eyes. "He's insisting I make out my applications for university. He says it used to cost lots of money and I should be grateful for that now it's free and use it."

Haroon stayed silent. He never knew what to say when Yoshi was like this. He picked up his fork and took a bite of the poutine he'd bought for breakfast.

"I mean," Yoshi went on, "I don't understand why I should be forced to do something just because of the way it used to be. I don't even know what I'd get a degree in."

"Maybe he figures you're smart enough to do one," Haroon said, scraping up gravy. "You would be a great bioengineer, and there's a shortage of those."

"Pfft," Yoshi waved it away. "Bioengineering is hard work. I want to go flick around in Japan for a while. Check out Korea, Singapore. Try out at some tournaments."

Haroon nodded. Yoshi was an avid Outrider player and would do well on the pro circuit. "You know what your dad thinks of e-sports, though."

"*Shakai ni kōken shinai*," Yoshi said bitterly. "Does not contribute to society. But it does. Entertaining other people with your skills is a totally valid contribution." Yoshi went quiet for a while. "So what does old man Subhan want you to do?"

Haroon thought of this morning, when his father had kicked him out of bed as he always did. "You get up, you go to work, you don't go wrong," he'd said in his thick accent. He was always telling Haroon not to go wrong. He never let on what he felt was right, though.

"I dunno. All he ever says is I'm to stay out of trouble and stay away from the gangs and especially the government. And I have to earn like a man. As long as I don't eat too much and keep out of his way, I don't think he cares."

Yoshi slumped forward, crossing his arms on the table and resting his chin on them. "Sorry, man. You must feel I'm a real dingus sometimes. I just wish I wasn't their only kid, eh? That all their attention wasn't on me. And I wouldn't have to help make up for how grandad went all *johatsu* on us."

Haroon reached out and patted him on the arm. Yoshi had recently confessed the reason they'd immigrated was that his grandfather had up and disappeared one night, becoming one of Japan's 'evaporated people.' No one understood why. Had he lost his job? Was he been cheating on his wife? The shame of his disappearance had prompted the family to move out of Japan. His grandmother had passed away not long after the transfer. That it had taken Yoshi this many years to tell him this showed how much it still disturbed his family, even now, in a new country.

Yoshi glanced down at Haroon's long sleeves. "Your dad ever work out you'd been tatted? Or that you're at school?"

Haroon pulled up his sleeve to reveal the digital ID Yoshi's father had helped him get a few years back. Given he was so young, the tattoo wasn't very complicated and it only extended a little way up from the top of his wrist bone. The first line, a solid, boring glyph, was his government-issued social insurance number, encrypted and machine-readable only on the underside of the wrist. A second pattern showed he'd completed his early education. He'd add his third, for high school graduation, soon. He couldn't wait to see that pattern on there: school had been a real struggle.

"I doubt it. He's never said, and I don't bring it up." He'd always suspected his father would be angry about it. Haroon had never worked out why they lived where they did. All he knew was he couldn't remember if he'd ever seen his father smile.

"So what will you do?" Yoshi asked him.

Haroon rubbed his ear and stretched. "Not sure. Well," he corrected himself, "I know the minute I qualify for the basic income and can afford it, I'm moving out of J." That was another thing that made him angry, learning that would be available to him at age of majority. He understood why the government didn't pay parents an extra income per child; that would have been too easy to abuse. But why hadn't his father signed up for basic for himself? It would have made his childhood so much less painful. "I'm not sure after that. I don't think I'd be any good doing any more schoolwork. It's like my brain evaporates when I step in the room and I have zero attention span. Kind of also depends on how things go with Saba."

"Ooooh, is it getting serious? Making life plans together?"

"I didn't say that. We've only just had our third date. But I like her enough I'd maybe want to stick around Toronto for a while. We'll see." He grinned. "I'll be your coach."

Yoshi swiped one of his fries. "Have you even played Outrider? That's sort of required, you know. To be a coach you have to play the game."

Haroon laughed. "No, I like the strategy games better. The fighting ones get too repetitive for me."

Yoshi's wristband chimed. "I gotta go. Comparative literature this morning." He faked an exaggerated yawn and rolled his eyes. He left Haroon to the remains of his breakfast.

Haroon tried to get back into his book but failed. His mind kept straying back to the question of what he would do. He struggled to think of a job either his father or Yoshi's father might approve of. Subhan's stern face loomed in his thoughts, and nothing came to mind.

RAY

He was bigger than his mother.

Because Mick had taught him how to break into the storeroom at the museum at Black Creek, and each of the grungy shops along Jane in turn, and to take just enough food from each that no one would bother calling the cops or to send anyone around to smash his knees. Mick had shown him where to stash the food and spare clothes so they wouldn't be stolen, and how to wear layers with the worst stuff on top so he didn't look like he was worth mugging. So he'd eaten. Regularly. And stayed warm. And gotten bigger.

He was bigger than his mother.

Ray didn't know why it had taken him so long to realise it. One day, he looked down at her, bunched his fists, and saw fear. In *her* eyes.

That was all he needed. There would be no more cowering in a corner during one of her drug-induced breakdowns.

He turned around and walked out. And walked and walked and walked for hours until he reached the edge of a lake that looked as big as a sea.

It occurred to him now in a sudden wave of shame and guilt that Mick hadn't disappeared.

He had.

When he had fallen to his knees at the lake that day, someone found him and told him he was in Woodbine, whatever that meant. And then they took him to a place with red doors.

The man who helped him clean up said Ray looked about fifteen. His name was Mubarak. He reminded Ray of Peter.

~

A chest so tight he could barely breathe. A heart that pounded so hard he was sure other people could hear it.

So much anger.

He was outside the District now. He totted up all the things they had denied him for years.

A roof that didn't leak in the rain and walls that didn't whistle and shudder when the December winds blew.

The knowledge that a stomach could be full *all day*.

Clothes that didn't stink of other people.

And the thingweb. A world of worlds within a world. There was so much there, he could see that. He wanted into it so bad he could taste it.

But he didn't have the tattoo that would let him access any of it, those inky smudges everyone was given at birth. All of them except him, of course.

He couldn't even read.

There were rules here. Stupid rules, meant to contain and confine. Ray had enough of confinement.

So much anger.

He raged at Mubarak for helping him live and he raged at Cedric for dying.

And he smashed things. So many things, they had to send him away.

He remembered Mubarak's sad face at the window.

~

Cold again — a deep, down-in-the marrow cold — and uncontrollable shivering.

Ray remembered bouncing from shelter to shelter in his fury until none of them would take him anymore and then freezing on the flows.

Maybe it had been the cold that put out the fire in his chest at last. But he found calm. Clarity.

He remembered swiping clothes from backyard lines. Ducking into the many gyms and fitness centres, making like he belonged there until someone figured out he didn't, watching how it was done, pouring his anger into beating down a pad or lifting chunks of iron. The reek of sweat. And showers. Soap. There was never enough soap.

Summer yard sales were great for nicking what he needed. Go early enough in the morning, and there were lots of crowds and no one paying attention. A pilfered antique primer on phonics. Teaching himself to read, slowly, painfully. A backpack. Memorising the recycling bot routes to rummage through everyone's bins the night before. He would have bought things if he could, but without access to the thingweb, there was no way to earn, and nothing he could do to pay. He'd learnt in the old days that they'd used physical currency, stuff you could hold in your hands. He wondered what it was like.

Peering through windows to see what the children were watching, to learn, until he became so absorbed he bumped the glass or made some other sound and freaked someone out. He learnt if he visited a restaurant and just asked for water, and it wasn't busy enough that they needed the space, they'd mostly let him hang around. Why wouldn't they? He wasn't causing trouble.

Yet there were the police, always watching, suspicious of his lack of ID, trying to herd him back into J-District. Or if they caught him, just dumping him there.

No way. Never again. He did not want to go back.

Had it really been that cold? He couldn't stop shivering.

KEL

The weights of the lat pull-down machine clanged down too hard, echoing inside the workout booth, nearly deafening her. She sat for a few minutes, breathing fast from the set she'd just completed.

It had been too long since she'd done any weight training. Or any serious workouts at all, really. She cursed having to exercise as a waste of time. Heck, she cursed sleep as a waste too. But lately, she'd been getting too little of either and her mood had been suffering for it.

Yesterday hadn't helped. They'd had their third meeting on the potential budget cuts and gotten no further ahead on how to deal with them. The meeting had devolved, again, into a shouting match over whose projects to prioritise and protect. Then she'd spent most of the afternoon being pulled into discussions dissecting the gathering and listening to speculation on silly things, like whether the herpetologists had a private protection agreement with Robert, or if the botanists would form a coalition against the ornithologists. It was as though all of them had forgotten it was an ecosystem and that one group couldn't do without the other to both maintain and study it. Kel had come home with a splitting headache, with nothing to show for the day, and gone straight to bed. This morning, she'd decided she couldn't face another day of squabbling, used one of her personal time allowances, and went to her gym instead. And anyway, her DPA had been scolding her about poor self-care for weeks. It was time to shut it up.

Someone banged on her booth. "Hey, are you done in there? Some of us want to work out."

She opened the door and stepped out. It closed automatically behind her and started its ten-second sterilisation cycle. The man waiting his turn threw up his hands dramatically at the additional delay. He was enormous. He obviously used muscle augs; his chest was unnaturally broad and his lat and trap muscles bulged against his shirt. Kel watched with amusement as he squeezed into the booth. He didn't close the door, instead making a big deal out of manually changing the weight stack so she could note it was at the highest setting. He then groaned loudly with every rep. She rolled her eyes and looked for her next station. She pegged him for a superhero fanman and idly wondered which character he was sculpting for. Hulk, maybe? No, she was sure Hulk was green, and this man wasn't tinted. Yet, at least. What had the other big one been called? Juggernaut?

She found the rowing machine booth and stepped in, making herself comfortable while her DPA interfaced with it, adjusted the tension, and set the timer automatically.

"What locale and time period would you like?" the DPA asked her.

Nothing with people in it, she thought. "Random location. Cambrian era."

The booth walls flickered and then vanished as they were replaced by a virtual reality simulation of a body of water. The rowing machine itself went into soft suspension mode, allowing it to bounce and bob like a boat would when she shifted her weight. She couldn't see land anywhere; when she looked down where the floor had been, the clear water teemed with life. The booth cooled.

She started rowing and got into a rhythm, doing long, slow strokes, and letting the motion ease her mind. She rowed off the arguments over money, the stress over her macaques, the paper that just wasn't coming together and the comment from her mother about a person not being able to make a difference; that one was still stuck in her craw. Kel had never wanted her own epitaph to read: I led a comfy life and contributed nothing.

The movement felt so good that she kept going after the timer beeped. The peaceful sensation reminded her of the time when she'd written a ten-thousand-word chunk of her dissertation in a single day because everything just flowed, and it had seemed like her brain actually felt warm afterwards. She thought about how amazing it would be if she could do that at will—

Kel let go of the rower handle so suddenly it arced away and thumped against the booth and she almost fell off the machine. A group of trilobites scattered at the sudden movement.

Being able to access the state of flow *at will*. How much more would she get done if she could focus like that again? How much could anyone accomplish? Her mind soared at the possibilities, the problems people could tackle if they could work well whenever they pleased.

Because of course, that physical and emotional state was simply the result of a specific set of physiological conditions: some neurons stimulated, others inhibited. Heart rate decelerated, theta waves midline, alpha waves over the temporoparietal region of the brain. Tibetan monks could achieve certain brain states at will, but it took years of practice because they were working from the outside in. Trying to shut out the hundreds of mental and physical distractions bombarding you was incredibly difficult, Kel knew. And there were billions of neurons to coordinate. But if you could record a few moments of the exact state of a brain and body that was experiencing flow, and then 'play' it back into the brain when you wanted it, on a loop... She could modify the implant she was using for the macaques; she'd already seen evidence of being able to send inputs...

Kel frowned. "Computer," she said. She'd never named her DPA, thinking it a silly habit to personify a program. "Why hasn't there been more work done in cognitive amplification?" Surely if she'd thought of this, someone else had? She described what she was thinking about.

Her DPA paused. "There may be at least two factors to account for a lack of progress in this area. In the late teens, several US-based start-ups selling products involving concepts like trans-cranial direct-current stimulation, binaural inputs, brainwave entrainment and the so-called quantified mind went bankrupt when it was revealed they had falsified data on the benefits of the products. There was also at least one major crowdfunding campaign for a home-use functional near-infrared spectroscopy device that was also a scam. Research grant applications for work in this field declined significantly during this period, as academics did not wish to tarnish their reputations."

"And the second factor?" Kel prompted.

"Concurrent with this was a crackdown by several drug enforcement agencies, in the last major campaigns of the so-called war on drugs, on a group of pharmaceuticals collectively known as nootropics. These were pharmaceuticals intended for other therapies but were also taken for off-label uses such as alertness, memory retention, focus and wakefulness. They became very trendy in the early twenties, particularly as concerns over artificial intelligences like myself rose. A rash of overdoses in South Korean college students brought the drugs to the attention of the media; subsequently, an unbeaten team in a high-level League of Legends e-sports tournament in China and several underage children in national spelling bee competitions were discovered 'brain doping' to win."

That was all Kel needed to hear; she wasn't planning to use any drugs. She burst out of her workout booth and headed for her lab.

MEIKE

It was late. Meike walked quietly to the lab door to look in. Kel was there again, working and muttering to herself. There was a schematic of what seemed to be an implant on the screen; it looked like Kel was changing it.

Satisfied she wouldn't be noticed, Meike moved on to the rear of the building, where the pharmacy was located. The locked door opened for her automatically, and the lights came on. The door closed softly behind her.

A woman who'd bought her a drink at Smashers, one of Fa's many strange friends, had found out where Meike worked. She had said she was in the market for 'whatever looked interesting at her lab.' Meike had a good idea what that meant.

The first and most obvious stop was the medicine cabinet, the contents of which she was now examining. There was plenty of high-grade ketamine, which would get a decent price. It was expensive to license to fabricate, so the stolen stuff could be bought at a hefty discount and still be profitable. She had taken some before, and no one had noticed. Fa would want to try some, too. And the gabapentin... yes, that could be worth trying to sell, depending on who the woman's contacts were. She searched through the remaining shelves. The only other thing they had significant quantities of was acepromazine. She doubted it would fetch quite the same amount as the other two, but some was better than none. She took a few minutes to access the inventory database to change some numbers.

Meike then had a look at the locked drawers under the workbench. There were controlled chemicals, but she had no interest in selling chemicals to make things explode. Because they weren't meant for human or animal consumption, they also probably had added tracers to identify the point of origin, and there were very few people who had access to this room. If she sold any of that material, it would mean police interest.

She surveyed the room. Most of the expensive-looking lab equipment had been open-sourced years ago or could be fabbed, so wouldn't have much value. They would also be missed. Her eyes lit on the lab fabber itself. Fab recipes? She didn't know how to crack the rights management on them, but she knew she had rights to transfer them once. Perhaps she'd offer to do that and let her contact work out the details later.

She heard movement somewhere, so she quickly tidied up and ordered the lights off. The door locked behind her as she went back up the hallway.

Kel was still working away in the semi-gloom. Meike watched her thoughtfully for a while and then went home.

HAROON

Subhan kicked him awake. "We go, five minutes," he said while Haroon scrambled around for his clothes, blinking. "I show you place for job."

His father was waiting for him at the front door of their apartment. Haroon shrugged into his ragged winter coat, one he'd accepted from Yoshi that was almost a size too small. His tuque and gloves were still damp from when he had come home in the blizzard last night, but he put them on anyway. They left their small apartment without saying another word.

It was so, so cold outside. The tiny hairs on the inside of his nose froze and prickled, and the wind knifed at his face. He hunched forward, jammed his hands deep into his coat pockets and hustled. It wasn't long before the tops of his thighs went numb underneath the thin fabric of his trousers.

Haroon glanced at his father, stomping along beside him. His red and puffy eyes were streaming. His nose, thick, lumpy, and crimson, leaked into his stiff, grey moustache. He didn't know why his father didn't let either of them grow a proper beard to cut the cold.

As they neared the edge of the District, an insistent beeping noise behind them made Subhan curse, and he stepped aside. A small snowbot wobbled merrily past, its large front blade scraping the sidewalk and funnelling last night's snow into a belly lined with heating strips. It reached the end of the flow, paused, and then urinated meltwater into a storm drain, before turning back for another pass of the other half of the sidewalk. Subhan looked as though he might kick it but thought better of it. Police vehicles always circulated this part of the city; you never knew, with their tinted windows, whether an officer was actually in one, ready to dash out and pick you up.

They walked for about half an hour before they arrived, thoroughly frozen, at the pickup point. His father grunted a greeting at the other workers who were already there waiting. Mercifully, a mass transit pod pulled up promptly, and Subhan shoved his way to the front of the line to get on, stamping his feet as much to warm them as to get the snow off. Haroon sat beside him, blowing into his hands. His fingers ached.

The pod accelerated the moment the door closed, slowly at first, to allow everyone to get a seat, and then quickly, pressing them all back into their chairs. The display at the front said it would take them into Vaughn, and the waste-transfer station at the site of the old Keele Valley landfill, in about ten minutes. Haroon sighed. He was too young for the work here; he knew from the odd time his father had spoken to him about his job that they only took workers aged twenty-one and up. Haroon had hoped he'd be able to escape the district before being brought here; he hadn't counted on his father lying about his age.

A short, muscular man with a glossy black ponytail

bumped past Haroon and rather forcibly pushed Subhan over to take a seat next to him. He glanced over his shoulder, noticed Haroon, and stuck out a hand. His grin was feral. "Tomasso," he said, in a thick Sicilian accent. He had a scar through his right eyebrow.

"Haroon."

"Subhan, my friend. Your boy? Don't deny, I can see you look the same. You been holding out. Tsk, tsk. I didna know you had family. When you gonna get into that mech suit? You've gotta provide."

Haroon was amazed to see his father tense up. "Who let you here?" Subhan growled. "I do not want mech suit. I know how goes."

"You've seen how it works, then you know how useful it can be, yes?" Tomasso replied, his accent drawing the word yes into two syllables. "Getting borebots unclogged and shoring up tunnel walls is hot and dangerous, no? How long you wanna stay a charity case?"

Subhan clenched his fists, and Haroon suppressed a gasp. Tomasso wasn't the first person to suggest the workers at the Keele landfill mining and reclamation project were only there because manual labourers were pitied. Nor would he be the first to say the workers were there simply because they were cheaper than the specialist bots required for the job. The words burned Haroon even so.

"I do not want mech suit," Subhan said. "The rent starts cheap and stays cheap until I fat and happy. Then it going up and up and up. Soon? You have half my packet. No, no, no. I know how goes."

Tomasso shook his head. "Nonsense. Rumours. Lies."

"Yeah, well, Bob Kennedy," Subhan said, and then, to Haroon's horror, he went pale.

The smile on Tomasso face froze. "And just what you think you know about my friend Bob, eh?"

There was a long, tense silence. "Bob... he uh, killed self," his father said at last, "because he no pay his bills."

Tomasso relaxed just a little. "Yes, he did. But Bob, so he no manage his money right? Should no prevent you from making a smart decision. In fact, is good reminder that tragedy can strike anyone at any time." He patted Subhan's shoulder firmly and then glanced back at Haroon. "For any reason. You think it over, yes?" Tomasso left to go have a chat with someone else on the pod.

Haroon shrank back into his seat and pretended like he'd been staring out the window the whole time. But it was too late. He could see, when Subhan turned to look at him, the humiliation, and the fury in his eyes.

KEL

Kel let out a loud whoop of glee and danced around the lab. Aadi hooted back at her and shook the bars of the enclosure.

She laughed at him and handed him a new toy. "That's my good boy. I'll let you out of there soon, I promise."

The macaque hooted again as she went to her desk to look at her data. For the first time in a long while, she felt excited about what she was doing.

She had designed some simple proof-of-concept tests. After spending several days tweaking an implant, she had taken Aadi — a macaque who was not yet part of her original study but a resident of the habitat — and installed it in him. She observed Aadi in the habitat for two days to make sure there were no issues with the installation, and then she returned him to the hospital area and kept him in the spot usually reserved for injured and recuperating wildlife. She let him be for another couple of days to establish a new baseline. Then, earlier this week, she'd fed him a substantial meal and recorded a few seconds of the brain and body signals for satiation as he sat around looking full.

For her first test, Kel set the brain patterns to play back on a continuous loop and observed him for several hours. Aadi continued to act as though he was full, lounging more than swinging and playing, and ignoring the food she had left out for him. She stopped the implant replay and watched him become more animated and then devour the food he had only just passed up.

After a few days of regular feeding and play in his enclosure, she let him go without food for a day. When she was sure he was good and hungry, she made a new recording of his hungry states and fed him another substantial meal. Ten minutes after he'd finished gorging, she put the 'hungry' recording on a playback loop into his implant and watched him fall upon an another serving of food like he hadn't eaten in days. No sooner had he done, then he searched for more to eat. She turned it off then, not wanting him to get agitated or eat himself sick. She'd run the numbers from the data log, to double check that what she'd seen in person were actually changes in behaviour. They were.

Kel did another little jig. It wasn't a proper study, of course. For that, she'd need to use more macaques, divide them into a control and a test group, and use a double-blind protocol. The study would have to be approved by the ethics committee. And there wasn't any evidence she'd be able to reproduce the effect with more complex states. Hunger was primal stuff.

But it was a start. A glorious start.

Her mind whirled as a long, branching to-do list unfolded in her imagination. How would she do this work alongside her other project and do both well? What other experiments could she design to test other body/brain state recordings? Kel wondered if it would work in a group environment or only on isolated animals. How would she set up the safety protocols so that the animals wouldn't be injured or unduly stressed? Would she need more money or could this be done within the confines of the current budget?

A new idea stilled her happy fidgeting. What if this had plausible commercial applications that could stave off budget cuts? Kel frowned. With the advent of custom nutrition profiles and home fabbers, the diet-product industry had died years ago, as few people had non-medical weight management issues anymore. She couldn't see any immediate benefit to tricking the body into feeling well fed. But other states of being? It wasn't what she had set out to study, but if it kept her main line of research alive…

"Come to think of it…!" she said aloud, making the macaque startle. "Computer, what are the university's regulations on intellectual property and patents? What is our policy on commercialisation?"

She sat down at her desk to listen.

SETH

It was bright and sunny outside, but Seth barely registered this as he came out onto the front step of his house, boyishly excited. A well-used delivery crate was waiting for him.

He heaved the crate inside and let the embedded scanner read his wrist. It popped open to reveal a stack of books. There were ten books each from the fiction and nonfiction bestseller lists. Seth had thought about ordering virtual copies but wanted to get a feel for how they were assembled and packaged. Plus, he fancied he'd use them for décor later: an inspiration bookshelf. The local megafab station had printed them up just minutes after he'd ordered them earlier this morning. He could smell the fresh ink.

Seth unpacked the crate, cleared the wreckage of his nephews' most recent visit away from his favourite chair, stacked the books there, and tossed the crate back outside for pickup and reuse. Then he made tea, and settled in.

He spent the day reading them, growing increasingly confused as the hours went by. At the top of the list was a novel about a divorcee who killed her ex-husband. There was no motivation ever given for the murder. Her character was flat and unlikeable, and the story didn't tie anything up at the end, leaving the reader hanging.

And how was this other book about a virtual reality star getting any good reviews? The writing was terrible, and the book was so poorly organised he thought it was a satire.

At least two that ranked well on three different lists were by someone named Mat Jameson. Seth looked at his publishing history and was sure he must be an AI. How could anyone produce so much in such a short time? Granted, not everyone had as many family demands as Seth did, but that kind of output seemed impossible. And why were *all* of them selling so well?

He made dozens of notes, looking at plots, characters, covers, even bindings and paper types.

Seth finished the day feeling bewildered. He had been expecting to find works that were simply much better than his own. And since Seth was no snob, so he'd expected a variety of things on the lists, from literary fiction to thrillers. He liked escapist fare just like the next person although he probably felt guiltier about consuming it than most.

But there were no common elements here that he could see, and while some were superb, others were objectively awful. Some books had great covers, some of them had terrible covers. There were new names and well-known authors on the list.

He went to bed, and tossed and turned for half of Sunday night, trying, but unable to find a pattern in the books he'd purchased. By 3:00 a.m., he gave up attempting to sleep and grudgingly activated his DPA.

"Tasha, I hate to ask you this, but… recall the books I bought earlier today. Review the industry trade publications and rumour sites. Are there any common factors in this list of books? Besides their list status? Or better yet, common factors that would contribute to their bestseller status?"

It took Tasha less than three minutes to come up with an answer. "All of them have marketing budgets at least one standard deviation above the norm. Would you like to discuss?"

Seth puffed air out of his cheeks in frustration. "No. Yes. I suppose."

"Discussion mode, on. Opening question: why did you wish to know this information?"

"Because I'm trying to find out why these books are outselling mine. I don't need a bestseller, I'm not naïve enough to think they'd hit those sort of lists given their subject matter. But my books are all but disappearing soon after launch, and I'd like at least mid-level sales, you know? Or steady low long term sales."

"Follow up question: why didn't you ask me to retrieve this information when you first purchased the books?"

"I wanted to review them myself, to see if I could figure it out."

"Follow up question: why did you want to do that yourself?"

Seth rubbed a hand across his chest. This is why he rarely engaged in discussions with Tasha: it always felt like a therapy session.

"I guess because it felt like cheating to use a tool… to use you to find the answer."

Tasha went silent. Seth walked into the kitchen to get some water.

"I cannot parse your last statement. You said it would be cheating, yet you are surrounded by tools you use every day."

Seth opened, and then closed his mouth. That was true, of course, and obviously so if he thought about it. "It's… it's different with art. Art is one of those things that separate us from the animals." He didn't add he also thought it was supposed to separate humanity from the bots. Then he felt slightly foolish for worrying about offending Tasha.

"So you think it is cheating to use tools for art?"

"Well, no," Seth said, thinking of the hammer and chisel his sculptor character was using, but also his own decrepit workstation. "But…" But what? He sipped his water. Were newer tools cheating while older tools were okay? Being called out on his irrational divisions — by his DPA no less — made him slightly grumpy. "You sound like my sister now," he grumbled.

"How do I sound like your sister?" Tasha asked.

"She's a philosophy major," Seth replied, taking another drink, reminded of a conversation he'd had with her months ago. "And she might say the greatest difference between humans and animals is that we can choose to optimise almost any aspect of our existence. We can have extended minds by offloading our cognitive burdens into tools like you, Tasha, and extended bodies with all our sensors, and augmentations, and mods, and such like." He yawned and considered his carefully programmed fabber. "She'd probably also call me a bloody fool for optimising my diet but not my career."

"I am sorry, I do not understand your—"

"It's all about the numbers, Tasha," he said, thinking aloud now. "Apparently, I need a job to make money to market my books whether I can keep my publisher or whether I go out on my own." He tapped his wristband and examined his calendar. An appointment to take his cousin to hospital for surgery. Another to accompany a niece on some university campus tours to help her choose where she wanted to go. And yet another with an uncle that wanted his advice on some investments because of the research Seth had done for a book. He grimaced. On some days, it was hard not to wonder whether he was so popular with his family just because he was unmarried and had no kids of his own. The fact that none of them seemed clear on what he did for a living didn't help. "And I now need to figure out how to boost my writing time because I'll have even less of it."

"I recommend getting some sleep."

"Thanks, Tasha. I didn't need you to tell me that."

MAURA

"Champagne?"

Maura turned her attention away from the two women who had cornered her and smiled politely at the young man. "Lovely, thank you."

While he served her companions, she scanned the well-dressed crowd, looking for a familiar face so she could make her excuses and escape. It was standing room only in the conservatory at Casa Loma, a brilliant white room with large arched windows dotted with frozen raindrops from the ice storm earlier in the day. The pink- and purple-hued rays of the setting sun made them sparkle almost as colourfully as the stained-glass dome above them.

"Such a beautiful venue," the woman who had introduced herself as Eve was saying. "I had no idea there was a castle right in the middle of Toronto."

"You must go on the walking tour," said her friend, Vania. "Especially down to the stables and carriage house. They have cars there that must be two hundred years old."

"Not quite," Maura said, wondering how the woman had missed the dates on the display. "Early 1900s, just like the castle."

"Just think," said Eve, who sounded to Maura like she

used this line a lot at parties, "there was a time when actual humans drove those things all over the place at high speed. Crashes were so common it was the custom to set up little temporary memorials where people died."

Vania shuddered delicately. "I cannot imagine. How could you possibly do that safely if you were tired? Or had some of these?" She held up her glass.

Maura cursed her bad luck. Small talk with strangers was the reason she avoided these charity events, and these two had glommed onto her almost as soon as she'd walked into the reception. "Indeed. That's why the bar and restaurant industry has boomed. No one has to worry about drunk driving these days." They gave her a blank look. "Look up the term. So, what brings you out this evening?" she said, hoping to steer the conversation to something more interesting.

Eve smiled. "Ah, Vania and I have recently moved here from New York, and we want to get involved in the community. You know, to give back to a city that has been so welcoming already."

Maura nodded. "The Toronto Symphony Orchestra is a good place to start, although, as you can see today, it already has a lot of support. If you're looking to help, there are quite a few others causes that aren't as well subscribed."

"Actually," Vania said archly, "we're here tonight because we want to start something new."

Eve took her cue from Vania. "Yes, we are going to set up," she paused dramatically, "a Toronto foundation for virtual reality addiction." They glanced at each other smugly and then at her, expecting an angry reaction.

Maura sighed and made a note to herself to chat with the organisers about the evening's guest list and their vetting process. Small talk was one thing; amateur hour was quite another. She drained her champagne and signalled for another. "As an immigrant once myself, I can appreciate being somewhere unfamiliar and not knowing the lay of the land," she said, as a waiter came up. She put her empty glass on the tray and took a filled one, holding their gazes the whole time. "But do you know who I am?"

"Of course," Vania said. "Yes, you're Maura Torres of—"

"So the plan was what, ambush me with your ideas, capture my disbelieving and angry reaction, and put it out on the thingweb feeds? Maybe encourage people to caption it?" Maura indicated a broach on Eve's dress. "I've seen that model in the self-defence catalogue. The red stone covers and disguises the camera lens, but the software filters the image so it looks normal and not all red. Pretty clever, modestly priced. Holds lots of video in memory. Or do you have it on a live broadcast now?"

Eve blushed. Vania looked embarrassed but defiant.

"I know that catalogue well because I shop it myself. It has excellent anti-kidnap technology for when I'm travelling, for example. I have a similar device, so I have my own records when dealing with journos or people like yourselves. You're not the first, you realise." Maura took another healthy sip of her drink, thinking that Yorkshire champagne had been one of the very few positives of the climate chaos of the last few decades. "Rule number one in activism," she said, "is to know your target. Did you really think you were going to shock me? Dear me. I started the national foundation for alternate-realities addiction. And I deliberately open-sourced the gamification methods the industry uses to keep players hooked so consumers can make informed choices. I was the one who pushed for industry standards on VR timeouts after four hours of continuous use. Thirty seconds spent consulting with your DPA and you'd be aware of this. So either you're incompetent or you're not really concerned about your cause and are only in it for the giggles and attention."

"Fine," Vania said, after a moment or two of guilty silence. "But I bet you hate the new VR tax!"

Maura didn't bother to disguise her scorn. "I was the one who proposed it to the minister! I'm not blind to the side effects of VR. Every new tech like this has unintended consequences, and we should work to mitigate them. Even if it wasn't the right thing to do, it makes good business sense not to harm your customers." She took another long pull of her champagne. "Rule number two of activism, by the way? Just making something for the feeds for people to point and laugh at isn't really activism. 'Raising awareness' sounds great, but should only amount to about five per cent of your budget, unless it's intended to be a fundraising campaign."

Just then, her wristband pinged, and she shooed the women off attend to it, grateful for the chance to get away.

It was a message from Pauline. She read it twice. It wasn't good news.

In fact, it could ruin everything.

RAY

Spring. Not so long ago.

Strolling along Philosopher's Walk, still in awe of the university buildings around him, wondering if he would ever get a chance to attend.

He'd made a deal: an apartment, no questions asked about his lack of ID, if he did all building maintenance for free. Ray had no idea if he could even do that kind of work, but he'd put up a convincing front. He could learn, somehow.

He had a plan: he'd get a proper job, once he was no longer someone of 'no fixed address.' To show he deserved integration. An application and the pain of a tattoo. The tattoo. That was the pain he yearned for.

His Digital Identity. To belong at last.

And then: school. Real school. Fabber pattern design. That looked interesting, creative, and a good way to earn a living.

And her.

That sweet-fierce woman he had seen with the long hair pulled up in a coil, who hurried through campus, not seeing, but thinking, planning, absorbing. He knew about that.

To be good enough for her.

HAROON

Haroon paced up and down the grimy, dimly lit kitchen, tearing off hunks of naan, shoving them into his mouth and swallowing them without tasting them. He checked the clock for the fourth time in the past ten minutes. Subhan should be home soon.

He had spent days with the career counsellor at school, taking tests, talking, poring over the options, trying to find a job he thought he could do. A way to earn a living, so he wouldn't have to dig out the landfill or depend on the basic income. Something that might make Subhan look at him a little differently. Like a man. Not a boy. Not a mouth to feed.

Today was the day to tell his father what he'd picked.

Haroon heard erratic thumping down the hall and hastily tidied up the breadcrumbs on his shirt. Soon after, Subhan slammed open the door. He stood there, weaving and blinking in the light of their tiny living room. His clothes were filthy from work.

The smell of alcohol reached Haroon's nose. A standard Tuesday night. He hoped it wasn't rye, as that always gave Subhan a vicious headache.

"Hello," he said, pleased his voice didn't shake. "Did you eat?"

Subhan blinked again and grunted at him. He came into the kitchen and sat with a thump at the kitchen table to be served. Haroon ladled out a heap of korma into a bowl and laid the bread on top. He gave it to his father and then got himself a portion, careful not to make it bigger than the one he'd just handed over. He sat at the table, and they ate in silence.

When at last Subhan pushed back his bowl, Haroon couldn't wait any longer.

"So, ah…" Haroon cleared his throat, and then the words came gushing out. "I've been thinking a lot about what I should do. You know. To earn. I wanted to pick something I figured I could be great at. And I thought about how it would have to be something I really cared about. Just so I could be good at it. And then I remembered how this neighbourhood really bothers me and how no one ever comes in here to clean it up, and so that's perfect, and I registered for a police academy today. It's risky work, but I think I'd be good at it. And it pays well. So it would be good. Yeah. To earn."

Haroon stopped, suddenly aware that was the longest thing he'd ever said to his father; and he was looking down at his bowl instead of at Subhan as he'd intended. He glanced up, hoping for a nod, and maybe a grunt of acknowledgement, if not approval.

Subhan was staring at him, immobile, a white-hot fury pouring off his body, the likes of which Haroon had never seen. When he spoke, it was with a deathly quiet voice.

"The police."

"Y-yes?" Haroon said. "The RCMP."

He stood, now completely sober, and slowly and deliberately spat on the floor by Haroon's feet.

"You. Get. Out. Don't. Come. Back."

He turned his back on Haroon and walked out of the apartment.

Haroon felt his arms and legs go weak, deflated, as though something had stabbed him, and everything had drained through the gaping hole.

He sat there for a very long time, not reacting. Not understanding.

KEL

Kel shivered and wished she'd taken her usual route through the habitat to warm up more quickly. She hated going out in the late February air, especially after eating dinner, as that always made it feel worse. But she'd needed to eat, and she had been fed up with the options available in the lab. Kel had treated herself to a nice big bowl of bánh phở and headed back to work.

She was anxious to return. She'd built three more prototype implants with an improved design, better pattern filters, and had drafted a presentation she planned to give to Robert. She needed to compose an application to the ethics committee to go with it before she went home for the weekend.

Kel got to the lab door and then stopped.

The outside door was wide open. It normally closed and locked automatically after anyone with an authorised entry signature in their digital tattoo opened it. Worse, the inside door was also open. Was pathogen scanning operational? Had someone come in without being scanned? It might not be too bad if they stayed in the office area, they could quarantine and scrub that, but contamination of the habitat could be disastrous.

It was still dark in the lab. Had the door malfunctioned? Was there something wrong with the facility controls?

She approached the door cautiously. "Hello? Robert? Bao-Yu?"

Did something just move? Her heart started thumping.

Kel stepped inside the door. The lights did not come on as they should have.

She cursed, and her mouth and throat went dry and tight. Everything seemed unnaturally quiet. It all felt very wrong.

Were the animals okay? Were they even still in the habitat? Or was that door open, too?

Something hard smashed into her shin and she cried out in shock and pain, falling forward. Someone grunted in the blackness right beside her — so close! — and moved and the hard thing came down on her head.

Then there was nothing.

RAY

A drone dropped out of nowhere, stopping between them. It was big and silent and unmarked.

It flicked a scanner beam, long and red, like a tongue, up the length of Mick.

And then it exploded.

Flashes. Mick blown backward, collapsing into a pile of shredded flesh. Blinding white light. Blood everywhere.

The emergency vehicles. Lights. Voices. A rough boot kicking his ankle. "This one is still alive!"

Too much pain to feel. Blackness…

That picture of Mick. Poor Mick.

Swimming, swimming upward, must get out, struggling, surfacing…

Ray gasped and blinked.

Everything seemed so dazzling. His eyes watered. He raised a stiff and shaky arm to wipe them and discovered tubes plugged into the back of his hand. He stared at them, uncomprehending for several minutes. Then he became conscious of tubes up his nose, and he snorted and gagged. He ripped them out. He let his hand drop, drained by the effort. A light flashed above his head. He strained to look around and figure out where he was.

"Easy there, luv." A woman with a brusque and confident manner came into the room a moment later. "You won't like it if I have to put those back in. Go slow and we'll see how we get on without them. All right then?"

Ray tried to nod. The woman seemed startled. "Oh my, we are doing better today, aren't we?" She came closer, peering at him. "You've been floating in and out of consciousness for a few days now since we took you out of the coma." She ran a gentle hand up from his forehead and over his head. Ray realised he had no hair.

The woman seemed to understand the widening of his eyes and gave him a soft smile. "Sorry, duck. Don't worry, we'll let it grow back soon." She frowned at some devices where the light had flashed. She glanced down at him again. "I'm Jane, by the way. And never mind if you can't remember that the next time you come to."

He tried to say his name. His voice was a hoarse wet growl. Jane clucked her tongue and patted him. "There'll be none of that just now, yeah? Save your energy." She turned away from him and inspected what Ray registered was an IV drip. A hospital. It wasn't all a dream — a nightmare — then. It had happened.

But he would not be deterred. "R-ay," he croaked.

Jane cocked an eyebrow at him, considered, and said, "Right, so you're one of those. Stubborn blokes. Hang on." She left, but returned quickly with a small cup of ice chips. She gently pushed one into his mouth.

It was cold and clean and beautiful. He closed his eyes, groaning a little. When he opened them, she was looking at him kindly.

"Where?" he whispered.

"You're at Toronto General," she said. "And you probably wouldn't be alive if you hadn't been so close. Do you remember any of it?"

Ray hesitated and then nodded slightly.

"Aye, well," Jane perched on the edge of the bed. "I'll not judge, and anyway it seems to me with the stuff you've been shouting in your sleep that you're not the type to be involved in any of this anarchist nonsense we hear about down south… I'm sure the police will still want to speak to you when you're ready though."

Ray was tugging at her sleeve, trying to get her attention. "H-ow… how long?"

"How long have you been here, you mean?" She pursed her lips. "That's a good question. They've had you in a coma since before I rotated in, so…" She consulted a device. "A couple of months, looks like. They took a whole bucket of shrapnel out of you, printed a new spleen, reconstructed your eardrum, loads of grafts, put you in the physio machine several times…"

But Ray had stopped breathing.

Two months.

Two months was long enough for the building owner to have given up on him. To have sold off everything he had so carefully collected and given away his space. Long enough for the job interview to have been marked as 'no show.' For him to have been labelled a skip. Not stable.

Not worthy of integration.

Marked.

He had lost everything he had been working towards.

Everything.

The low keening noise he had started making grew louder. The nurse halted her monologue and looked at him in alarm.

"Now what are you on about? You'll be up and about before you know it and you've got all four limbs, which is nothing short of a miracle, and — oh, lawks…"

The anger and helplessness of his teenage years came crashing back. He wanted to break something.

Break everything.

Adrenaline slammed into him and the heart monitor went wild. He sat bolt upright, ripped the drip out of his hand, threw it at Jane and roared, the cords of his neck bulging and his face flaming red. The fire in his chest consumed him. He stagger-slid out of the bed and smashed his fist into the digital chart above his bed. The glass shattered everywhere.

He was still reeling unsteadily around the room, tears pouring down his face, and screaming when a hospital security guard tackled him to the floor.

PART II

KEL

When Kel came to, it took her several minutes to realise she must be in a hospital. Still groggy and confused, it was several more minutes before she remembered why she was in one. She recalled being struck twice. What else had happened? She pressed the button to call for a nurse.

A few minutes later, a big man with red hair and a face full of freckles strode in. "And there ye be," he said, smiling. "My name is Sam." He checked the screen at the head of her bed. "And you're… Kel, it says here."

Kel thought she could detect a Newfoundland accent. He projected a friendly sort of confidence; she liked him right away. "Hi, Sam," she said. "Yes, I'm Kel. What happened?"

"Well good, glad we have the right person in the assigned bed," he winked. "They brought you in here with a wicked coupla knocks to the head and a right shin that has seen better days. You're a hockey player then?"

Kel laughed, despite her fears. "No, I wouldn't know one end of a puck from another."

"Just as well you don't play, as pucks really don't have ends, seeing as they're round and all," Sam replied, still reading her chart. "The short answer to your question is: not very much in the grand scheme of things, and you're goin' to be fine."

Kel tried to shift in the bed and winced when a pain shot up her leg.

"Yes, don't wiggle yourself about just yet. You've got a badly broken shin there, and the painkiller will only give you so much of a buzz. We've had to insert calcium phosphate putty, and you've got magnesium alloy pins and plates in there to make everything sit tight. Once the swelling comes down more, we'll print you up a cast for it, too. Your next interface with a food fabber should tell it not to supplement for magnesium as you won't need any spare when those bits start to biodegrade, but watch for that, eh?"

Kel nodded.

"As for your head, you took a solid smack to the back of your noggin, and then, I'm sorry to say, you seem to have fallen on your face. You've a right shiner, I must say. This your kit?" Sam pointed off to her left.

Kel gently twisted to look. Her bag was sitting on the night table by the bed. "Yes, that's mine. How did that get here?"

Sam referred to her chart again. "Says here one of your colleagues, someone called Bao-Yu? She put it in the ambulance pod. She found you. Is there a mirror in there I can get?"

"Yes, I have a silver compact in there. It was my grandmother's."

"Sweet," Sam said as he fished it out of the back. He pried it open and flipped it over to show her the reflection. One side of her face was swollen and the area around the eye was a deep purple.

Kel raised a hand to touch the bruise. "Ouch!"

Sam nodded. "Yeah, I know. We can give you something to speed up the healing there, if you like, or you can just let it fade on its own." He lifted the sheet covering her leg and checked it over for signs of infection. Satisfied, he tapped in some notes into her chart.

"There's… nothing else then? I wasn't… "Kel paused, unsure how to ask what she wanted to know. She'd been unconscious. "I wasn't assaulted in any other way?"

"Cor, no, there'd be more than just me here to see you if that were the case."

Kel breathed a sigh of relief.

"You'll get on the go in a day or two, looks like. If you feel up to it, the police would like to chat with you. Oh, and we've eight messages from your mother for you to call her when you're conscious."

"Eight!" Kel groaned. "How long have I been in here?"

"Only since last night."

Kel groaned again. Her mother was her emergency contact, so she would have been notified almost immediately. "Who should I talk to first then?"

Sam grinned at her. "Well, your mother was threatening to visit if she didn't hear soon…"

Kel reached for her wristband. "Say no more."

~

By the time Kel had reassured her mother she would be okay, half an hour had passed. A delivery bot trundled up to her bed, bearing a tray of food for lunch. As hungry as she was,

Kel had a hard time choking down the tasteless sandwich and the tea that tasted like ditch water. When she was finished, she was certain hospitals sourced their food fabbers from the lowest bidders.

There was a knock at the door. A constable from the Toronto Police Service nodded at her when she looked up. "Are you Kel Rafferty?"

"Yes," Kel said, and he came in. He introduced himself as Gobinder Singh. He pulled a chair over to her bedside, carefully hitched up his trousers and sat down. Without preamble, he put a small tablet between them and thumbed a control. It flashed to show it was recording. Then he said, "I understand you were involved in an incident last night? Please tell me what happened."

"I'm not sure I can tell you much," Kel said. "I came back to my office after going out for dinner and found the doors open. I stepped in to see what was happening, something smashed into my leg, and then I guess I was hit on the head. That's it." She gasped as she suddenly recalled her concerns going into the dark room. "The animals! Was the habitat breached?

Singh shook his head no. "It doesn't look like it. I interviewed your supervisor this morning and reviewed security camera footage, and nothing appears to be missing from the... habitat, I guess you called it. So we can probably rule out animal rights activists. Was anyone at your office with you or were you working alone?"

Kel felt limp with relief. She wasn't sure she could handle another research setback. "Alone, as far as I know." The constable gave her a quizzical look. "I mean, people pop in and out of there all the time, but there wasn't anyone with me."

"And you weren't expecting anybody?"

Kel shook her head. "No."

"Do you often work alone at night on a Friday night?"

Kel wasn't sure she liked the tone of his question. He made it sound as though it was a weird thing to do. "Quite a bit, I guess, yes. Why?"

"Is this a habit, I mean? Part of your routine?"

"I… don't really have a routine. I work some nights… but not all nights, if that's what you mean."

The constable stroked his beard. "Thank you for clarifying. Did you have anything of value on your person that is now missing?"

Kel thought about the contents of her bag, which usually included a hairbrush, some cosmetics, and perhaps a snack bar or two. "No, nothing. I mean, I didn't have anything really valuable. I haven't checked whether anything was taken."

"How about in your office? Was anything of value there? From the footage I referenced earlier, it looks like whoever hit you took things from your desk."

"No, I—" Kel gasped.

Two of her prototypes, a printout of the schematics and a hard copy of her proposal had been in her desk.

"Are you sure?" she said. "My desk?"

"Yes, one of your colleagues," and here he swiped away the recording screen on his tablet temporarily to check his notes, "Robert? Robert said it was your desk. What was there, Dr Rafferty?"

Kel didn't know how to answer. Who would steal what she was working on? Who even knew what she was doing? She had told no one at the office yet. Would saying something now end up putting details into the public record? How would that affect the university's chances of getting a patent? Kel hadn't wanted to say anything to Robert about the side project until she had a solid presentation put together. She might not know a lot about office politics but she knew enough to want to avoid presenting something half-baked.

"Dr Rafferty?"

"There would have been… parts. Parts of an implant for the macaques I work with," Kel replied.

"Do you know the registration number of the implants?"

"Uh, no."

"Would you have the registration number on file?"

"No, they were parts, as I said…"

Constable Singh nodded, the badge on the front of his turban flashing. "So nothing of value then?"

"Well, it's really rather important I get those back."

A flicker of annoyance went across his face. "Do you have an IP address for them? Are they GPS-enabled? RFID?"

"No, I'm afraid they weren't hooked up into the thingweb, and as I said, they were parts so—"

"Dr Rafferty, I appreciate you might be feeling a bit rough, but I will need more than this to do anything further. Were these parts expensive to license and print? Were these exclusive licences? Are you or the university going to be significantly out of pocket here, or what?"

Kel felt like a stuck audio loop. "No, they were parts, but the configuration I was using was unique and—"

Singh raised his hand. "Okay, look, I get it. They were important to you. Do you have any idea who might have taken them? Anyone have a grudge against you at the moment? Does anyone have access to the contents of your desk or to your work computer, or know what you were working on? Is this an intellectual property concern?"

She reviewed what had been happening lately. If he wasn't impressed by a real robbery, he would not think much of dead lab animals or her missing data, given she couldn't prove anything about either. All she had were coincidences. Reluctantly, she shook her head.

Constable Singh looked disappointed and stopped his tablet from recording any further. "I think we're done here," he said, standing up and giving his tunic a short, sharp tug to straighten it. "I'm sure you will remember your unique configuration, and you'll print new parts. But I can't do anything if you won't tell me more. If you think of anything else I should know, please get in touch." He left.

Kel was suddenly conscious of how much her leg was throbbing and her head ached. She felt miserable. All she desired was to close her eyes and have it all go away. But one thought kept repeating in her mind: What would someone — someone willing to attack a stranger — do with a device that recorded and replayed brain states?

RAY

March had decided to come in like a lion. The weather was absolutely foul.

The day had started with howling winds and driving snow — big, wet clumps that landed with a splat and stuck to everything. By noon, it had warmed enough to change into freezing rain, and now the footpaths were slick with it. The pod flows had slowed to a crawl. Walking was treacherous. The cold was damp and phantom-like, getting under his skin, sucking the heat out of him and making him ache.

After another interview with the police that involved an apology from a senior officer of the force, they had discharged Ray from the hospital that morning. Nurse Jane had been by, bearing a coat and a set of clothes she claimed to have discovered in the lost-and-found closet. There was a hat in a pocket of the coat. Ray had nodded at her wordlessly, not trusting himself to speak to her lest he break down again. A social worker had also come by, handing him a printout of an address for a shelter that was expecting him, and another place for outpatient care. He'd left the paperwork on his hospital bed.

He was done with false hope.

Instead, he'd slipped out of the hospital, anonymous still, and half-walked, half-slid his way across the city, trying to keep his back to the wind as much as possible. Darkness had fallen by the time he made it to Santoro's Bar. When he'd lived in this area, it was a place he avoided; everyone knew it was a gang stronghold. Today, it was the only place he could think of to be.

Santoro's was a tiny watering hole located in a shabby old plaza just off Sheppard. It was dim inside; what little illumination there was came from the yellowing lights in the ceiling, filtered through a cloud of cannabis and tobacco smoke. The smell gave Ray an instant headache, but at least it was warm and dry. The walls were lined with booth seats that had dull green upholstery, while the middle of the room was filled with low wooden captain's chairs and small tables.

At this time of day, the bar wasn't crowded, although it wasn't empty either. A couple of men sat near the window, older guys originally from the Mediterranean dressed in white shirts and dark slacks, the sort he'd seen before who seemed to have little to do all day but sit and watch and smoke and occasionally talk to each other in a language Ray didn't understand. They stared at him brazenly, looking at him go across the room. Some pothead drifters sat in a corner, laughing too loud and too often, and eating something greasy Ray couldn't quite see through the haze. There was one woman at the bar, drinking beer, and watching a hockey game on an ancient television screen. She was cursing at the hometown Leafs and shouting abuse at the goalie. Ray chose a stool at the far end, well away from her.

He pulled off his hat, as it was already dripping cold water down his neck in the warmth. The ice that had coated it cracked and fell off, shattering and skittering along the floor. The bartender, a woman with heavy makeup and a face that had seen too much sun, gave him a look and came over.

"Hey hon, you here just to make a mess of my floor or you gonna buy a drink?" Her voice sounded like she'd been inhaling the smoke in here for years and chasing it with scotch.

Ray nodded and reached inside his coat. By chance, he'd come across an open recycling bin behind a corporate office downtown and had found a pile of dead tech — old mobile phones, cracked tablets — just there for the taking. The waste disgusted him but was a bonanza he desperately needed. He'd grabbed everything he could and stashed all but one phone in Mick's old hideaway by the culvert on the way here. He slid the phone across the bar now.

She picked it up, gauged its heft, and gave it a look-over to see if the case was still intact, or if it had been stripped of its precious elements. Ray knew it hadn't been as he'd already examined it himself. Finally, she nodded. "That's good for a fair few," she said. "Mind, it's worth what *he* says it is, okay? I'll set you up a tab. Food or drink?"

"Food. I'll take a plate now. Give me your cheapest special. And the name is Ray." She tossed a towel at him and took the phone away, disappearing into a room behind the bar. He mopped up the ice chips. After a few minutes, she returned with a steaming bowl of borlotti beans and rice.

Then she leaned back against the counter behind her and crossed her arms, not bothering to disguise the fact she was looking him over, assessing whether he'd be trouble.

He pulled off his coat and draped it dry-side up on the bar stool before sitting down again. He tucked into his food, knowing he must look a sight with his hair still not grown in yet, a livid scar up the side of his left cheek and another slashing through his right eyebrow. There were similar marks elsewhere on his body. The hospital had done what they had to by law to make him healthy again, including physiotherapy and giving him a few packets of pain drugs to take away. But they didn't do any cosmetic work as he wasn't registered under the province's health plan. Perhaps he would have gotten that later, had he gone to the shelter. It was too late to think about it now.

He was also skinny, pale, and weaker than he realised. His legs were throbbing, and his feet felt like they were smouldering. Everything was stiffening, and now he had stopped walking, he wasn't sure he could stand up a second time.

"I haven't seen you in here before," she said after he'd eaten nearly half of the meal. It tasted so good, especially after his long, cold walk. She brought him a glass of water, which he drained immediately. "You new to the neighbourhood?"

"Old," Ray said between bites. "Was here before, went away, came back."

"Hmmm," she said. "I'd have stayed away."

Ray shrugged, trying to remain nonchalant. "It didn't work out."

"I can see that. So what now?"

"I dunno," Ray mumbled, chewing. "Find a job, I guess. Find a place to stay. Right now, just eating is good."

She seemed to take pity on him and brought him some hot cider, unbidden. Someone else approached the bar, and she went to serve him. Ray noticed the other patron knew her by name, calling her Sylvie. A regular, then. He looked around. A place like this wouldn't get too much in the way of random, drop-in traffic. It was likely almost all regulars here, which was why he'd gotten the examination.

Exhaustion hit him like a ton of bricks; he swayed on his stool. He'd exerted more energy today than he'd done for weeks. Ray wondered sleepily whether he'd said enough to Sylvie, or if it had been too much and too obvious. Was everything as it had been before?

He hoped so. Because unless something had changed, the Twins controlled the bar. And the Twins were his key to getting involved in J-District's crime scene.

There, he'd have access to the tools he needed to hack into the system.

He would find out who had killed Mick.

MAURA

"I'm not sure I understand," said Pauline. "What's the significance of this set of queries? I mean, when I flagged these for you, I thought they would be of interest but not a major concern. Forgive me for saying so, but you seem quite upset."

They were in Maura's office, sitting across from each other — Maura at her desk, and Pauline seated in a chair in front of the desk. Maura was reading and rereading the report on the screen. She gave Pauline a half-smile.

"You were right to follow your instincts," Maura said. "And this is why I have human personal assistants here at the office. As good as the AIs are, they still don't make the broad leaps of pattern recognition and understanding. You saw something here that was interesting, even if you weren't one hundred percent sure why." Maura shut down her computer, feeling troubled. "Did I explain to you why we intercept and monitor traffic to the patent sites on the thingweb?"

"No," Pauline replied, shifting in the seat to get more comfortable. "But I assume, based on the summary reports that land in my email, that we're watching for filings from competitors and for a technology of interest in order to acquire it?"

"Yes, we do that. However, if that were the only thing we were looking out for, we'd only be tracking a narrow set of queries and a mundane kind of traffic to the site. We're also watching for patents that can disrupt the VR industry altogether."

"I still don't follow," Pauline said, looking puzzled. "These queries came from a neuroscience lab that works on, from what I can tell based on the publicly available research papers, the pathology of Alzheimer's disease. The queries seem to be focused on methods for recording brain states. When I highlighted these, I was wondering if we could use whatever they were developing to get an edge on improving the VR experience. You know, like what Sensate was trying to do, only not with something random, like odours. If we could see exactly what the brain was doing during a VR session and compared states between people who aren't negatively affected to it to those who are, we might see what the difference was."

"Oh," Maura turned her computer back on and made a few notes. "Actually, that's fantastic. I hadn't considered that. My focus was on the brain state replay queries."

"Why? What's the big deal about them?" Pauline asked.

"Don't you see?" Maura said, leaning forward. "If I could get a recording of your brain and play it back in me? It's the ultimate in voyeurism. Wouldn't you love to know what it was like to feel like someone else?"

To Maura's surprise, Pauline blushed. She looked to be profoundly embarrassed by the thought. But her voice, when she spoke, betrayed none of what she seemed to be feeling.

"Okay, but still, what's that got to do with us?"

"Because VR is all about vicarious experiences," Maura said, getting up to pace. "Simulations. Even those universes where we've lavished huge amounts of attention to detail, like the individual leaves on the trees. It is one person's best attempt — usually the VR storywriter — to convey what it's like to experience something. The story can be something the author has actually experienced in real life, but they still need to translate it in a way you or I can experience it. And no two people interpret a VR session the same way. It is, by definition, virtual. These brain recordings would be the real thing." Maura stopped pacing and turned to face Pauline. "If this tech got released to the public… everyone would try to hack into everyone else to take a recording. And it wouldn't stop with humans, I bet. Animals, too. Everybody would want to know what Fido felt like fetching the ball."

"What is it like to be a bat?" Pauline mused, half to herself.

"What?"

"Nothing, sorry, just thinking of my undergrad philosophy courses." Pauline said, and then her eyes widened. "Okay, I get it now. VR would seem like old news."

"Precisely. The market would evaporate overnight."

"Would it really, though?" said Pauline. "The patent site queries were very, very basic. There's nothing else out there like this on the market, so far as I can tell. That suggests this is early in the game and the tech is primitive."

Maura sat down again. "It is now. But there are two things to keep in mind. One is that technology can advance quickly if it has the right people working on it. Where some inventions languish for years, others take off like a rocket. It all comes down to backing and marketing. Two is that if this person at…" she frowned at her computer, "this University of Toronto affiliate, has thought of something like this, then someone else has something similar, and we don't know about it yet. The history of science has shown this quite a few times — lots of people having the same idea at roughly the same time. And Pauline, there are eight billion of us in the world, at least half of which are smarter than you and me."

"So what do we do?"

Maura clenched her jaw. "We get control of this technology, come hell or high water. Find out who ran those queries and let's dig up everything we can find on him or her."

HAROON

Haroon's head was spinning.

He was in a stuffy meeting room in the Cadillac Fairview tower, listening to a recruiter talk about how to join the Royal Canadian Mounted Police. It was way more complicated than he thought it would be.

He would have to have an official hearing test, and an official vision test; he would need to go through a battery of psychological assessments, and he would have a physical check-up, too. They would look into his past, establishing whether he had a criminal record. And that was just the first stage! Haroon considered the physical training requirements the recruiter had on display at the front of the room. Just to complete the cadet training — assuming he was even invited to become a cadet — he would have to complete an obstacle course, a push/pull test and a weight-carry test. Haroon looked down at his body. He was reasonably fit, but he was no athlete. It was looking like he would have to train five days a week, for at least eight weeks, just to get up to speed.

And then, the academic part. Cadet training would take him twenty-six weeks at the academy, which meant lots of *intensive* studying. He'd had a hard enough time concentrating throughout high school, much less in a pressure cooker. It would also mean leaving Toronto for a while. He had never been outside Toronto; the thought scared him. A lot.

And if he passed all of that… he'd be working four days a week and taking university-level courses one day a week, since undergraduate degrees were the minimum requirement for staying in the force these days.

Haroon squirmed in his seat. Now he understood what Yoshi meant when he complained about the bioengineering program being hard work. This seemed overwhelming.

The recruiter finished his presentation and forwarded the application form to everyone's wristband. A few people surged forward to ask the recruiter questions. Haroon nodded a polite thanks to the officer and left the room.

Out on the flow, it was evening, with the sun kissing the horizon. It was getting hazy, and there would be fog tonight as the night chased away the tenuous spring warmth. It should have felt great, being out here right now, standing in full view of everyone walking by or whizzing by in pods, and not having to slink through the shadows. Instead, everything felt… weird. Like it was all about to go upside down.

He hadn't told Yoshi or Saba he'd been kicked out of his home. He wasn't sure why. Maybe because it was because he knew Yoshi's family would offer him a bed and he was tired of being a charity case. He was tired of being *their* charity case. And he sure didn't want Saba to think he was some homeless loser. Instead, he took a chance on an apartment search and found a tiny place that would take him in now because he was due to qualify for the basic income next month, when he turned eighteen.

Haroon tapped his wristband to call for a pod to go home and then pulled his hands up into his sleeves to keep them warm. His bed. His kitchen. At first, he was giddy with the idea. When the landlord left him alone, he'd run around the whole apartment like a madman, for the sheer joy of racing about in his own space. He ordered a bed and some bedding from the megafab station and assembled it with his own hands. He fantasised about how he could decorate the place and planned where he'd squeeze in a desk and his very own reader, and he debated what he'd call his personal assistant. Last Sunday, Haroon had spent the entire afternoon browsing in the animal companions shop, wondering if he was ready to apply for one. It seemed like a daunting responsibility.

Then he'd run up hard against the limits of his budget if he wanted to eat and travel in the city at all for the next month. And there was no one else in the apartment with him. Subhan had never been home very often — if he wasn't working, he was drinking — but at least there had been the comfort of knowing it was a shared space. That there was someone you were at least vaguely connected to in the same orbit. He thought about Saba and wondered how far along someone had to be into a relationship before they considered living together.

Someone walking by bumped his shoulder. Haroon popped his hands out of the sleeves, instinctively ready to defend himself. But the person simply mumbled an apology and moved on without giving him a second glance.

It was so different out here.

And he was very much alone in the crowd.

RAY

Ray awoke to stabbing pain.

He blinked, looking around in confusion. He was on his side, on a bench, and there was a wooden table in front of his face.

Ray tried sitting up, gasped as what felt like lightning shot down his left leg. He grabbed at it, nearly falling off the bench to the floor.

"Tsk, tsk," said a voice. "Sciatic nerve, probably. Hurts like a sonofabitch when it's pinched."

Ray groaned and pulled himself upright as gently as he could, bracing one elbow on the table and the other on the top of the bench.

The bar.

He was still in the bar and had been lying on a booth seat. Morning sunlight streamed into the windows at the front of the bar. He had never felt so stiff and sore in his life. Another shot of lightning seared up his leg, forcing him to stifle a gasp.

In front of Ray, a man had pulled over a chair and put one foot up on it, leaning forward with one elbow braced on his raised knee, his hand casually holding a long cigar. The smoke curled upward in the silence. His posture was at once relaxed and yet coiled, tensed. He was lean, dressed in deep black trousers, an exquisitely tooled belt, and an ivory shirt. The sleeves were rolled to the elbows to reveal intricately tattooed forearms. His face was all angles and sharpness, and his black hair was wavy, short, and worn tucked behind his ears. He wore bright gold loops through each earlobe.

Ray mumbled apologies and moved to get up. The man made a single gesture, a simple 'no' movement with his free hand, and something told Ray he'd better obey. He subsided.

The man took a deep pull on his cigar, then exhaled the smoke in Ray's direction. "I come up here," he said quietly, "every day, to eat my breakfast, and generally do not find people sleeping in my restaurant. But Sylvie here tells me you passed out at the bar. You must have been very polite to her for her not to just dump you outside to freeze to death. You wouldn't have been the first."

Ray tried to speak, found he couldn't, cleared his throat, and tried again. "I suppose I was. No reason to be rude."

"How interesting," the man said. "There are lots of people who find they can be rude for no reason at all." He flicked ash off his cigar into an ashtray. "Why did you pass out? What are you using? Smack? Special K?"

"Nothing," Ray replied. "Just exhausted. Walked all day. Been sleeping rough. Hard to catch a good night in this weather."

The man squinted through the smoke at Ray, his black eyes drilling into him. "The other guy look as bad as you?" He indicated Ray's scars.

"Worse," said Ray, truthfully.

The man laughed. "My name is Dominic," he said, stubbing out his cigar. "You'll eat with me. Sylvie!" he called. Sylvie came out of the kitchen. "Two this morning." She nodded and disappeared again. Dominic gestured to the bar, and Ray hobbled over to it.

His stomach churned. Dominic! One of the Twins, here, now, already. He wasn't sure whether to be happy or terrified.

They sat at the bar, and Dominic turned to regard him closely. "Do you know why you woke up?"

"No?"

"Because I was staring at you. Took just under a minute. You felt me staring at you. Instinctively woke up. Don't you find that fascinating?"

Ray wasn't sure how to respond. "I guess so."

Sylvie appeared, bearing two plates piled with food. She set them down and left, to return with a jug of water and two glasses.

"Steak and chips," Dominic said, brandishing his knife and fork. "My favourite breakfast. *Mangia, mangia.*"

"Thank you," Ray said and meant it. It smelled good. He ate some chips, cut off a piece of the steak, and put it in his mouth.

"It's real, you know," Dominic said.

"I'm sorry?" Ray said, chewing.

"Real meat. Not factory grown," Dominic said. He sliced a large chunk of his steak and brought it to his nose to inhale its scent.

174

Ray stopped chewing. "Like, cut from an actual living animal?"

"Yes," Dominic replied, watching him. He took a drink of his water. "Very expensive, very hard to get these days. These animal rights *idiota*. They would have us forget we are animals, too. Deny our true nature."

Ray fought an overwhelming urge to gag. Nausea slammed into him. Carefully, slowly, struggling for control, he resumed chewing. He was sure he could taste blood. The blood of a *creature*. He swallowed.

"It's… interesting," he finally managed. "I've never had it before."

Dominic laughed again and raised his glass at Ray. "Clever enough not to claim it was delicious. Or maybe not clever, but just honest. Well done either way." He put down his glass, reached over with his fork, and speared Ray's meat. He slapped it down on top of his own steak; the sound of the two pieces of meat smacking wetly into each other made Ray want to vomit. He looked down, focusing hard on his chips.

"So, you are a man who is without a home and without a job," Dominic said. "And you appear to know how to survive a fight, at least. What shall we do with you?"

Ray reached for his water and was appalled to feel his hand trembling. He switched to picking up chips with his fingers instead, hoping that wouldn't offend. "No, nothing," Ray said, demurring. "I'm sorry I disturbed your breakfast. I didn't mean to be here this—"

"Do you know who I am?" Dominic asked him.

Somehow, Ray found the strength of will to look at Dominic and hold his gaze steadily. "No, I'm sorry. I don't," he lied.

"Mmm," Dominic said, cramming more meat in. It took him a few minutes to finish the mouthful. "Then you will learn," he said, finally. "I need people like you who are... situationally pliable. You will learn I am not someone who hears the word 'no' more than once from the same person. You will learn to like your new job. And..." he waved a forkful of steak at him, "you will learn to eat this and love it."

MEIKE

Meike and Fa were in her bed. He was naked, lying on his stomach, his face half-pressed into the pillow. She had pulled on blue stretch pants and was sitting up, trying to read some papers.

Fa stretched and yawned. "What is that? You've been staring at it for an hour," he mumbled.

"Something from work," she replied. She was reading Kel's draft presentation for the third time. She had worked out what the device did and sensed — dimly — it might be important, somehow. But she couldn't see the point. Other people seemed boring to her. Meike didn't think they'd be any more interesting on the inside than they were on the outside. Why would you want to record and replay people's brain states?

Fa rolled over, throwing an arm over his eyes.

"What would you do if you could record an experience and then play it back again?" she asked him.

A smile spread across his face. "That's easy, baby. I'd replay the day I met you. I wish I had thought to video it."

Meike grimaced. He had to go soon. She hated it when they got this way; she had no idea what they wanted from her and no way to give it.

"No, I meant the actual feelings," she said. "The exact thing you felt, physically and emotionally."

"Whoa," he said, sitting up on his elbows. His hair was a mess and his eyes were glassy and dilated. "That would be so blast. You could record the best sex you'd ever had. Just think of it! Getting your mind totally blown any time you wanted. Without having to work up any kind of sweat." He rolled on to his side and stroked her leg, grinning at her. "I'd put all of our sessions on heavy rotation." He flopped backward and sighed dramatically, still rubbing her. "That puts all kinds of questions into my head. You come up with the most profound stuff at the weirdest times."

She batted his hand away, irritated with him. He laughed and went to the bathroom, probably to look for more poppers.

Meike tucked the papers back into the drawer of the night table beside her bed. What would be the point of experiencing the same sensation and over again? It would be boring. He might not be wrong though. Although she couldn't recall of a single thing she had done that would be worth repeating once, much less several times, if Fa thought those would be good things to record, other people might as well. And they would pay good money for the privilege.

She contemplated using the device and opening a studio, like the mod shops across the flow. But the idea of having a constant stream of Fa-like people coming to her all day was unbearably tiresome. Might be best to sell the thing outright to someone who could see its potential.

She knew a few people who would appreciate a deal like

that. Better still, they'd want to pay in a cryptocurrency to remain untraceable, just like her. In fact, she could sell it three times over, as she had two prototypes, and the schematics. The question was, could she demand a higher price for exclusivity, or should she risk selling it to different people? Some of her contacts were pretty dangerous.

Meike heard Fa snorting something in the bathroom. She wondered if they would also scare him off.

KEL

Kel limped along Church St., gripping her bag tightly, the mech-assisted cast on her leg whirring almost imperceptibly with every step she took. The leg still ached like anything: a deep, down-in-the-bone ache like the one you got with a bad case of the new flu, but concentrated in the one area. But at least she was up and moving. She'd been going stir-crazy during the prescribed 'take it easy' period.

She turned the situation over and over in her mind. The police hadn't uncovered anything in their investigation of the assault. Whoever had done it had been astonishingly careful: dressed all in black, the attack done at night, and power to the lights and doors cut ahead of time, power backup also shut off. It rendered the security cameras almost useless, the footage only good for confirming the assailant had been human, and not, say, macaque. The only DNA found in the area belonged to herself and her colleagues, so he or she must have worn a superclean suit. But who? The question was driving her crazy.

And making her paranoid. She tucked her bag a little more snugly under her arm and picked up the pace. She still had her third prototype device — fortunately, it was the most recent iteration — but hadn't found a secure place to keep it. Clearly, it wasn't safe at work: the robbery had proved that already. Her apartment would be no safer. She'd thought of taking it to a bank safety-deposit box, but that would require registering and describing exactly what it was for insurance purposes. Kel had been keeping it in the bag she always took with her everywhere, but given how absentminded she could be when deep into her work, there was a good chance she'd end up leaving it in a pod, or at the library or something.

There was only one way she could think of to keep it with her at all times. Really with her.

She stopped at a shop painted bright purple. The sign said BIOHACKER BITZ. She took a deep breath and went in.

A cheery young woman greeted her. She'd shaved herself bald and had devil-horn implants above her eyebrows. The sight of them gave Kel some comfort; the protrusions looked as though they were growing under the woman's skin like natural bone would and had always been there. If this was the quality of the work they did here, she was a great advertisement for them.

"Can I help you?" she asked.

"I hope so. I need something installed. A nurse at the hospital said this might be the place to get it done."

"We get quite a few unofficial referrals from the hospital, actually," said the woman. "Of course, that's because most of the staff work here."

"I'd hoped that was the case. But I don't understand why this stuff isn't done in a real hospital."

The woman smiled. "I'm not going to hit you with a sales pitch, but let me assure you, this is a real facility, every bit as professional and sterile as the hospital. The work isn't done where you were because it's not regulated. Legislation hasn't caught up to reality, as usual. So that creates a liability issue for the hospital that it isn't willing to take on."

"But you are?"

"Yes," she said simply. "The doctors and nurses who form the corporation here have hefty insurance premiums. But they want to do the work, so here we are. What is it you're looking for? Piercings? Genital beads? Something like this?" She indicated her head. "I gotta tell you, there was a guy in here last week that had some pretty wild headwork done. Klingon-style head plate. It's gonna look blast when it's healed."

"Nothing that stylish, I'm afraid," Kel said, trying not to be distracted by thoughts of how she'd look all bald and bumpy. She paused for a second, mustering her courage. "I just need a standard brainjack."

"Nice," the woman replied. "Come back into the guest gallery, and I'll show you some of the different ports available. There will be quite a few waivers to sign. And we must discuss payment."

The woman disappeared from view. Kel stood there, irresolute. Bao-Yu was right: she'd never considered one before now because she didn't really need one. Between her work ethic, her natural ability, and coffee, she'd always been able to achieve what she set out to do. Augments took time and money she didn't want to give up. Kel had never gone around announcing as much but on the odd occasion when someone had asked her what she'd had done, her answer had always resulted in an awkward, strained silence, and a false smile. "Oh, lucky you," was the usual rejoinder.

Kel frowned to herself. Was she afraid? Brainjacks were fairly commonplace these days, and complications were rarely reported. And she had designed the implant for the jack! She knew exactly how it functioned. A memory of her grandmother, looking confused and helpless, rose unbidden in her mind. The surgery wasn't entirely risk-free: she could suffer brain damage if she was unlucky or the surgeon was having a bad day. Surgeries were still not one hundred percent complication-free.

But she needed to keep this implant hidden while she tried to work out who had stolen the other ones and why. And she had to test and improve the design, on an accelerated schedule now, to get a patent secured and to ensure the device was properly licensed and used the way she had meant it: to capture and replay flow. There wouldn't be time for extended studies and animal experiments until after she'd safeguarded the rights. She would have to be the test subject.

Kel squared her shoulders. She recalled the long line of researchers in history who had experimented on themselves to prove their work, like Werner Forssman and his cardiac catheterisation or Barry Marshall and his ideas on what caused ulcers. This device was too important, it had too much potential, for her to get squeamish now.

She followed the woman into the gallery.

RAY

Ray's heart was pounding. It was close to three in the morning.

He was about fifty metres away from the administration office for the Mississauga megafab station and warehouse. It was humming with activity.

Even from this distance, he heard the fabbers working through the night in the giant building behind the office. This station made pods on demand for the city and the surrounding area. At one end of the building, transport trucks automatically delivered worn-out pods for disassembly and recycling by the bots inside. Through a window over in the middle portion of the building, Ray could see a machine working on constructing the main body of a pod, the print head zipping back and forth, adding layer after layer of material to build up the lightweight shell. Elsewhere in the building, other parts were being fabricated and advanced down a conveyor belt, where they were gripped, manoeuvred, and assembled into a full pod by a small, silent army of robots. Completed pods drove themselves out the other end to their destinations. There were no humans around. There wouldn't be until at least 9:00 a.m., and perhaps not even then: the technicians travelled a circuit of megafab stations and checked each one only intermittently or when there was a self-reported issue to fix.

Ray was crouching in the bushes with Tomasso, Dominic's lieutenant and the man who'd drawn the straw to break in the new guy. A short, thickset Italian with long black hair swept straight back into a ponytail, he had classically Sicilian features and sad eyes.

"Tell me again," Tomasso said.

"I walk up to the building like I'm meant to be here," Ray said. He wore a dark blue technician's uniform. "I palm this device against the door." He raised a small black box that would wirelessly retrieve a master digital key signal from the software controlling the lock. Originally intended to be on a programmable master key, the manufacturer had carelessly left it in the lock itself. There were hundreds of these locking systems throughout the city, and most of them remained unpatched. "If it doesn't work, I pretend there's a problem with the door itself. I fiddle with it and walk away, tapping my wristband." Ray touched the fake wristband on his wrist.

"And if it works?" Tomasso asked him, before turning his head, hawking, and spitting into the bushes.

"Then I run the plan."

"Okay, go." Tomasso made himself more comfortable. "You get caught, I leave. I've never seen you."

"Got it," Ray replied.

He backed out of the bushes and headed for a section of the parking lot where they had determined the facility cameras didn't quite cover. It would look like the pod had parked at the far edge of the lot and he was walking in. He tried to walk as naturally as he could, but he was vibrating with tension. Ray hoped the camera resolution wasn't good enough to pick up on that.

He made it to the office door and, in one smooth motion he had practised dozens of times, let the black box drop out of his sleeve and into his hand as he raised his wrist for a scan. Ray waited, his breath coming fast now, for several long seconds, before he heard the welcome sound of a bolt clicking back. The door swung open for him.

He glanced around, trying to get his bearings. Tomasso had made him watch intercepted security footage over and over dozens of times to learn the routine of the technicians here, and to figure out the layout of the building, but it had all been shot from above. It looked very different in real life, here on the ground. He consulted his fake wristband as though to remind himself what the trouble was supposed to be here. Then, gritting his teeth, he walked purposefully — he hoped — to the main computer.

Ray positioned his body between the screen and the camera he knew would be behind him, so it wouldn't be possible to see what was happening. Tomasso had walked him through what a tech would be doing. One, wake the screen. Two, tap for a login prompt. Pause for a beat. Three, let it identify you. Four, gesture for the trouble ticket screen to see if anything needed doing. Only Ray wouldn't be doing any of that. He would never actually wake the screen; instead, he would just be going through the motions. At the second step, waiting for the login, he casually shifted his weight and placed his left hand on the table next to his screen.

In his thumb, a micro transdermal transmitter executed a program that bypassed the facility's thingweb traffic router authentication mechanism. Within a few seconds, it had changed the admin settings for the domain name system to a malicious IP address owned by Dominic's gang. Ray sucked in his breath. The insert was cheap and not well made and it generated a lot of heat. He could feel it burning his thumb from the inside out. His mind flashed to the bomb and the shrapnel that had sliced into nearly every part of his body. He gripped the table hard, struggling to resist the urge to rip at his hand to get the device out.

After counting out the right amount of time in his head, he walked away from the computer, took a quick look around as if he was being sure nothing else was amiss, and left the building. He walked back to his ghost parking spot, his hand pressed hard against his hip to control the pain.

"Done," Ray panted. "You didn't tell me this was going to light me up so bad."

Tomasso smirked and held out his left hand to check Ray's thumb for a look. Without warning, he gripped Ray's hand hard, brought a knife out with his right hand in a paring grip, caught the raised edge of the transmitter and flick-ripped it up and out. Ray jerked and swallowed a shout. Blood welled up out of the hole. He clamped his other hand over the wound but his palm was so sweaty it made it burn all the more. He swore as loudly as he dared in the night.

Tomasso just laughed at him and slapped his cheek. "Not bad, not bad. Most new guys piss themselves before they even make it to the door." He flicked the bloodied transmitter into his shirt pocket and the knife disappeared as well.

Ray sucked at the wound, cursing some more. He took in Tomasso's relaxed expression and judged that he might have earned the right to ask a few questions now. "What did we even do in there?"

"Vampire exploit." Tomasso gestured for them to go back to their private pod, parked some metres away down the dark flow. "The transaction traffic here will be parsed and a small fraction of every purchase will be shaved off and deposited with us instead of the manufacturer. We have hundreds, maybe thousands of these going."

"Why did we have to go into the building? Why not just hack it remotely?"

Tomasso shook his head. "Easier to get inside and pretend to be a device on the internal network than to crash it from the outside. Besides, it gives us something to use for hazing." He clamped a hand on the back of Ray's neck and pulled him in for a rough hug. "You got nerve, I'll say that. You might stick."

Ray clapped him on the shoulder, acknowledging the compliment, wincing as his thumb twinged and bled some more. Two days in and he was a thief who was missing a chunk of his thumb.

He wondered just how much finding Mick's killer was going to cost him.

SETH

As it turned out, Seth's job interview at the Xperience Centre, nestled in the tech park in Toronto's waterfront, was first thing in the morning on the Ides of March. He tried not to read too much into that.

He stepped into the foyer that was painted all black and illuminated with ultraviolet light. His tan shirt glowed in the semi-gloom. It struck him as a bit tacky, but then again, he wasn't sure what he had expected. A virtual assistant materialised.

"Hello!" it said brightly. "May I please read your ID?"

Seth raised his hand so the VA could scan his wrist.

"Seth Bacchi," the VA said. "Welcome. Your interview will take place on the second floor, in the conference room. Please proceed to the left and use the lift."

"Thank you," said Seth. It was hard not to be reflexively polite when the assistants took on corporeal form.

He went up to the second floor and walked down the corridor until he found the right room. Even though he was early, someone was waiting for him already, seated in one of many chairs around a large table. She stood up when he came in.

"You must be Mr Bacchi," the woman said, standing and

bowing slightly. He bowed back, glad she wasn't one of these old-fashioned types that still insisted on handshakes.

"Just Seth is fine," he replied.

She directed him to a chair opposite her. "I'm Selena."

Once they were settled, she said, "I'd like to take a moment to thank you for your application to become a Dragon Slayer Farmer. Now, I know you have put what I'm about to ask you into the application form, but, can you tell me again if you've played DS before?"

Seth knew now, as he'd known when he had filled out the form, there was no point in trying to fib his way through the interview. This conference room would be loaded with sensors to detect his heart rate, his respiration, the moisture appearing on his skin, his eye movements, and gestures. "Not really, no. I'm interested in being a Farmer for a few reasons. First, because it would allow me to set my own hours. As a creative, this is important." She gave the thin, somewhat insincere smile of someone who has heard this response, or one very much like it, hundreds of times. "Second, because doing something in VR would disconnect me from my work for a while, and that will ultimately help me in my writing, keep me fresh. And of course, I want the money."

"You've had VR experience before, though?" she asked.

"Oh yes. Games-wise, I did the complete King's Quest series when it was ported to VR, for example. I've also done the tours of places like Chernobyl and McMurdo Station," Seth replied.

"That matches what you said. Thank you for verifying

your ownership of your application in real time. Now, I'm pleased to tell you your personality testing shows you're a good fit for the job."

Seth raised one eyebrow. "What do you look for, anyway? I'd have thought you'd test for more physical traits, like stamina and hand-eye coordination."

Selena laughed. "Well, you need to have a bit of that, and be fit, but honestly, we're far more interested in the mental aspects. For this job, at least, we like to screen for attitudes. For example, we don't want obsessives or bullies. We have enough trouble with that sort of thing with our paying customers, and it costs us a lot to police it. We want to make sure it's a safe and pleasant experience for all of our players, so we do not want to hire people who will cause problems."

"That makes sense, I suppose."

"Good," Selena said. "Now, how much do you know about our Farmer program?"

"I'm not clear on what they do, except play the game for pay," Seth replied.

"Okay then," Selena said. "A Farmer plays the game for someone else."

"Eh?" Seth said. "How does that work?"

Selena shifted in her seat. "Let's say you had already played Dragon Slayer as a Mage character, but then wanted to play again it as a Warrior, but you didn't want to go through all the tutorial quests and the skills-acquisition quests. These things take time. A Farmer plays the game to create a customised starter character, and then the buyer takes it over."

"Why doesn't Xperience just generate these for sale?"

"Human psychology," Selena sighed, as though it was the bane of her existence. "Game purists — our hard-core fans — would melt the discussion forums down if we offered preset characters. They say it allows rich players an edge, and it's cheating. And they're right, to a degree."

"But...?"

"But," Selena continued, "game farmers would exist anyway. Purists or not, there's a huge market for this kind of thing. There are other games in the industry where the farming companies for it set up in unregulated countries overseas, especially where there isn't a basic income yet. They hire players to farm, pay the workers a pittance, and keep them in the VR for much longer than is safe."

"So you set up your own farmers," Seth suddenly understood. "That explains why the application was to Dragon Rush, but I reported for the interview here."

"That's right," Selena said. "Dragon Rush is the 'officially tolerated' farm company, yes. It's our subsidiary. We don't hide that, but we don't advertise the fact either. People who want to level up fast, can, while Dragon Slayer doesn't look like it's allowing paid accounts a leg up and gamers aren't worked to death in the process."

"A polite fiction," Seth said, thinking it would make a great title for a book. "That's fascinating. I did not realise there was that kind of illicit ecosystem around the game." Ideas for plots flitted through his mind. Perhaps this wouldn't be so bad after all.

"You would be surprised how creative people can be

when it comes to hacking the system," Selena said, standing up. "Let's get you to a VR room and let you play an introductory round."

Seth followed her out of the conference room, and into a lift. "Gaming rooms start on the third floor," she said. "Support, design, management is all on the second floor. IT and all the primary servers are in the underground levels where it's cooler. We also use the heat from the servers to help warm the building in the winter."

They arrived on the third floor and stepped out into a nondescript hallway. There were doors every ten metres on either side. Most of the doors had red lights on them. She stopped next to one that had a green light, and they went in.

The room was empty and barren, save for an omnidirectional treadmill that took up most of the floor, and a clothes rack on the back of the door. Selena eyeballed him, measuring, and then selected a lightweight exoskeleton and haptic feedback jumpsuit from the rack. He put it on over his regular clothes, and then she handed him a pair of wraparound goggles that had blackout cuffs around the eyes. She showed him how to bring the earbuds out of the goggles.

Seth put the goggles on and popped the earbuds into his ears. He couldn't see anything, and Selena's voice was muted now. Selena guided him to the centre of the room, gave him the controller gloves, and wished him luck. He thought he heard the door close behind her.

Suddenly, his suit vibrated all over and he was transported… to a grassy meadow, set on a gentle hill on a beautiful, sunny day. He touched his head in wonder as he felt a gust of wind in his hair. A herd of sheep was running past him, bumping into his legs, bleating loudly. He looked in the direction they were running and saw there was a mediaeval-era village below, in a little valley. Smoke curled up from the chimney of one of the huts. It seemed rather quiet; he saw no one walking around the village although he would have expected people out and about, doing their chores for the day, especially since the sun was high in the sky. But overall, it seemed peaceful, even bucolic.

Then something made him turn around and look up.

A giant blue dragon was bearing down on him, fast.

He could hear its massive wings beating, thumping the air to keep its gigantic body aloft. Seth felt the wind again, this time against his face. It looked so real.

He saw it opening its mouth and drawing breath to expel flame.

It had very large teeth.

"Aw jeeze," he said, and ran.

KEL

Bao-Yu surprised her and gave her a hug when she returned to work. Robert shook her hand. Even Padraig allowed her a gruff nod to welcome her back.

Meike, however, was nowhere to be seen.

"She quit," Robert said. "Two days ago. No reason given. I've had a vet tech from the zoo doing cursory check-ins on your critters and she's on call if there's an injury or something. But we must get someone new in."

"Do we have the funding?" asked Kel. "Or could they could chalk that position up to our budget cut?" She didn't need the extra work, but she wouldn't mind not having to micromanage another assistant.

Robert astonished her by grinning. "Oh! I forgot you hadn't heard. No cuts needed. In fact, we've gotten confirmation we'll be getting an increase this year."

"But…" Kel couldn't help but think of all the time they'd squandered arguing about what to slash. "How did they go from wanting to cut to giving us more?"

"Politics," Robert said. "The opposition released their post-convention interim platform, and the government repositioned itself as the more progressive, pro-science party."

"You mean had the opposition come out with an aggressively progressive — is that even a phrase? — policy, the government would have cut even deeper?"

"Wow," Robert laughed. "That knock on the head must have turned on the proper cynical circuits up there. Now you're getting it."

He left her to her work. She popped into the hab with a bag of fruit, and was mobbed by the macaques almost immediately; if she hadn't known better, she would have said it seemed like they had missed her. Aadi climbed up to get a hug, so she laughed and cuddled him while she stroked another one clinging to her leg. After a while, they got too rough and inquisitive with her cast, so she tossed the remaining fruit a distance away and let them run after it. She returned to her desk to check on her datastreams and spent the next few hours polishing the final draft of the paper Robert was still expecting. It wasn't likely to be substantive, but she hoped being assaulted might at least give her leeway on that score.

After she sent it off, she pushed back from her desk to do some gentle stretching. For some reason, today her mind kept going over the conversation she'd had with the cop in the hospital. He had asked something about other people having access to her computer. She knew she hadn't talked to anyone about her project, so was hacking a possibility? Just how secure were the facility's machines?

Kel rolled her chair to her desk and checked the background processes of her computer. Nothing obviously wrong or out of place there. No software she didn't recognise. She shut down all the programs on her computer and checked again, this time watching for data traffic to and from her workstation. All seemed as quiet as it ought to be.

Could there be something elsewhere in the system? She would have to have a chat with the IT manager, Dan, if only to rule out the possibility of a hack. She got up and put on her coat, heading out the lab's back door to the little corner office across the quad that represented the entire IT department. Kel wondered how she would phrase her questions so she didn't sound paranoid or give away details of her project too early.

She found the office empty.

"You literally just missed him," someone said, behind her. One of the vet techs, walking by. "He said he was going out for a walk and a lunch break. Be back by one."

"Ah, thanks," Kel smiled. "I'll leave him a note."

She ducked into the office. Kel *had* only missed him by a minute or so, as his screen hadn't yet gone dark and locked down.

And that gave her an idea.

Kel poked her head out of the office to look in the hallway. It was empty. She went back in and closed the door.

She sat at the desk. His screen had an array of utilities running, all to automatically check for and run software updates, monitor traffic, performance, and log events. For the most part, the system administrated itself. Actually, it seemed to administer itself quite well, because there were also quite a few retro video games available on Dan's station. Kel wondered how often he manually looked at things.

She clicked open each tool one by one, scrutinising them for anything unusual. Although she'd never been interested enough to take it up as a career, she'd gone through a hacker-fascination phase as a teenager, reading up on the exploits of groups like NoSec and HackW0rmz, and getting involved in some hacking herself. The tools had evolved since then, there was enough that was familiar that she knew what she was looking at.

Kel checked the time. Ten minutes had elapsed already.

She reviewed the server monitors, the registries, and a few random logs. Nothing jumped out at her as unusual. Another fifteen minutes gone.

Kel found the network analyser and poked around. There was too much to get a handle on.

There were voices in the hallway. She froze, unsure how she'd explain what she was doing if Dan came in.

The voices faded away, and she let out the breath she was holding. Working quickly, Kel set up a profile for herself in the analyser that she could access from her desk. She picked a nondescript name for it, memorised the address for the utility on the network and shut everything she'd opened. Then she checked the hallway. Seeing no one, she beat it back to her office.

~

Kel poured herself another coffee and cursed. She rubbed her port and implant. Maybe she should make use of her new brainjack and buy additional augments for her memory.

She sat at her desk and tried logging into the analyser

again. It told her no such profile existed. She restarted her computer. When it came back up, she entered the address for the analyser and logged in one last time.

It worked. Kel did a *yes* fist-pumping motion, then stopped and frowned at her screen. It had worked... but... the address for the analyser looked different. She copied the one she had open, popped open another command screen and put the address in there. Then Kel went back to her original screen and checked the history of her login attempts.

It *was* different. By a single keystroke.

She flipped back and forth between the two screens. There were two network analysers in the system, with nearly identical configurations. One was the version she'd accessed in Dan's office, to which she'd added her profile. The other was sending data outside the network.

But where? It took Kel another couple of hours to trace it back through a complicated set of proxies, forking paths and dead ends before she found the end point: a company named EduTain. She tried to get into the receiving port, but the connection was abruptly terminated. Apparently, they had better security there than they did here.

Kel didn't know what EduTain was, but it obviously wouldn't be a good idea to look it up while she was in her office. But now she had more questions than answers. Was this company monitoring this facility or was this somehow a clumsy misconfiguration issue? If it was watching, whatever for? The name of the company didn't sound like it would be interested in Alzheimer's research.

And the biggest question of all: had it been scanning *her* computer?

Kel sipped the dregs her cold coffee. Maybe it was time for a company tour.

RAY

He was rich.

Ray blinked and looked at the balance again. He'd never seen that much money in his whole life.

"Here," Dominic said, handing him the small black fob. "Cryptowallet. Keep it on you at all times. Get a chain or something around your neck, like. Use your thumb and forefinger to squeeze the end to get the balance to display again. That thing is keyed to work only for you — and me — but don't be stupid enough to lose it anyway. Use it to buy what you need from other members of the organisation. In this place…" Dominic used both hands to indicate his office, and by extension, his mini-empire. "In here, you got everything. Out there, you got nothing. Remember that."

Ray nodded. He was getting a small fraction of the cash being syphoned off the place he'd help hit the other day, after it had been laundered and converted into the mob's own currency. He couldn't imagine how much must be coming into the organisation overall.

Dominic's office, which served as the organisation's headquarters, was in the basement of the plaza. In his personal space, at the centre of the area, he had an antique oak desk he liked to prop his feet on when he leaned back in a soft leather chair. There was a fully loaded bar in one corner, a large wardrobe in the other, filled with pristine white shirts. Dominic's desk faced the descending stairwell; no one could enter or exit without him seeing it. To his left, were the rooms for his lieutenants; further on from that, a room for low-level rookies like himself. They all had access to a full beer fridge, a vintage pool table, and a large bank of restored video arcade games. To his right, there was a door that always remained closed. Ray had only seen one person go in there: a 'guest' of the organisation, who went in accompanied by Tomasso and someone else Ray hadn't met yet. The guest had needed help walking out and looked ashen.

When Dominic dismissed him, Ray went out for a walk around the block. Dominic's mob was much bigger than Ray had realised. He had thought, when he signed up, he was getting involved with some local petty thugs and thieves, and he would learn about the larger players and go from there. But given the money and activity he had seen so far, Dominic was a significant player in the city, and maybe even beyond.

And so much had changed for Ray so quickly. He had a home now — a teeny, one-bedroom apartment with a kitchen that was a little bigger than a storage closet, but it was a roof over his head. It was free, provided by Dominic with hints there were better spaces available for people who earned for the organisation. Now he had money, too, and he could buy a change of clothes. He was concerned how he would keep track of who was who as he started digging for information about Mick. A tablet wouldn't be secure, but perhaps an old-fashioned notebook and a pencil, kept tucked into his coat. He wondered if they had wired his apartment with cameras and mics.

He crossed the flow. Then there were these walks. Somehow, without him saying, people in the neighbourhood knew who he was with, and they got out of his way. Even the two old-timers who passed their days in the restaurant only glanced at him now, averting their eyes if they saw him looking back. Sylvie, who had been friendly before, treated him with deference now.

Ray had gotten more from this group in a few short weeks than he had in an entire lifetime with his mother and the system outside.

A dog bounded out from behind a building, tongue lolling, panting hard. It ran up to him, its tail wagging furiously. He crouched down to stroke it, and it licked his face. Well-groomed and overweight, it had escaped from someone's yard far away from here. Ray noticed a small black box on its collar, covered in slush, a red light flashing. He cleaned it off and wrapped his fingers around it to warm it until the light turned green. The dog squirmed away, but Ray figured its owners would pick it up soon now it was connected to the thingweb again. Then he caught sight of his own bare wrists, and stood up quickly.

He should have been happy.

But he wasn't. Ray hadn't grown up dreaming of being a crook.

He walked faster, feeling the damp cold seeping into his core.

When he got back, he found Tomasso waiting for him.

"C'mere," he grunted.

They went back into the games room. There were two open bottles of beer on the card table. They sat down, and Tomasso handed him one. "Boss says you're the lowest rung on the ladder. Congrats."

Not sure what he meant by that, Ray clinked bottles with Tomasso uncertainly, and they both took a long drink of beer. They drank in silence, Tomasso being a man of few words.

Ray suddenly felt very lightheaded. He eyed the label of the bottle to see how strong it was. He wobbled and grabbed at the table.

Tomasso put his bottle down and gripped Ray's elbow.

"Nothing personal," Tomasso was saying. "Just part of the business."

~

Some days later, Ray was carefully taking the steps down to Dominic's office, his stomach churning. He had awakened about twelve hours after losing consciousness with a foul taste in his mouth, a splitting headache, and a bandage on the back of his head. He'd staggered into the bathroom to see what had happened, but all he could see was something hard and black at the base of his skull, crusted in dried blood. Ray went to his kitchen and discovered instructions to rest and then report to Dominic. There had been a steak in the fridge to eat. He hadn't touched it.

Dominic was at his desk, having a quiet conversation

with three of his men. They all stopped when Ray arrived. Dominic dismissed two of them and beamed.

"Ray, Ray, Ray, I knew I could count on you to show up again. How you feeling?"

"Okay," Ray said. "I—uh, I don't know what's going on."

"No, of course not," Dominic said. "We didn't tell you. We have something we would like you to help us with, and you didn't have a brainjack. So we had to fix that. Now come along."

A brainjack! Ray felt a little woozy at the thought someone had drilled into his head. He tried to blot out a vision of himself, out cold in a back alley somewhere, a grubby robot surgeon bearing down on him, bigger than the one he remembered—

Dominic gestured and Tomasso guided them both to the door that stayed closed. Ray was close to panic. What had he done wrong?

The door opened into a room that was empty save for two chairs and two small tables beside each chair. They were facing each other. In one chair, strapped in tight, was a young man of about twenty, looking miserable. He had a black eye and a split lip, and there was a strong tang of urine in the air. His left arm was bound to the table, with the hand palm down.

Dominic bade Ray sit in the other chair. Quaking, he did, expecting to be tied down. His mind was racing, struggling, trying to figure out what he might have said to offend someone in the organisation already. He couldn't think of anything; Ray hadn't even really poked around about Mick yet. He had done exactly as he had been told. Hadn't he? He pressed his lips together. He would not embarrass himself by asking stupid questions or grovelling.

"Now then, Ray," Dominic began, lighting up a cigar. "First, I should tell you what's about to happen here does not leave this room. Second, I should say that I have always wanted to bring our ways into the twenty-first century. I grew up seeing other *famigle* live and die with the drug trade. All that struggle, that strife, all of those guys who fell under the spell of a drug themselves, all of it suddenly worthless when the drugs were decriminalised. Who wants to buy a dime bag probably cut with flour off some mook on the flow when you can get high-grade goods at the local liquor store? Tsk, tsk. Such a waste of effort. Whole families, penniless and scrambling."

He dragged on his cigar, relishing the taste. "Extortion is still a good racket, though. We run a great line on mech suits for the chronically underemployed. And printing cut-rate parts for sale on the black market." He grinned. "But the thingweb, brother Ray, that's where it's at. The incident last month where the emergency services number went down? That was us. Oh, did we ever make bank with that."

Ray couldn't help himself. "The city paid the ransom?"

"Of course not," Dominic waved his cigar, looking scornful. "No, they paid one of our 'companies' to fix it. Which we could since we'd broken it to begin with. And we now have a standby fee in case it happens again. It won't be long before the contract is up for renewal." Dominic walked slowly, going behind the other man's chair, puffing smoke. The man hunched over, trying to resist the urge to look back. He flinched when Dominic leaned a hand on the chair.

"Unfortunately," Dominic continued. "We still have to do… this sort of thing. But this can get messy and…" Dominic wrenched the man's head backward suddenly, to look down on him, "it can leave a mark. I don't like evidence." He paused, and fixed Ray with a stare. "Did you know, Ray, I killed my sister so there wouldn't be a witness to one of our jobs?"

The revelation both shocked and didn't shock him. "I didn't know you had one," Ray lied again. At least now he knew what had become of Drea.

Dominic watched him for a moment longer and then nodded. "She was my twin, you know. Sometimes I think I can still feel her presence. A phantom twin," he grinned, and then waved his cigar as though this had all been distracting. "Anyway, today we will test something new." From his back pocket, he pulled a small device. "We were sold this by a very enterprising young woman. It will work on anyone with a standard port." He pushed the man's head down and inserted the device into his brainjack. "You see, Ray, it records brain states. You can take snapshots of what's going on, how the brain is responding and so on. And it allows you to replay it at will. Tomasso?"

Tomasso surged forward, pulling out a rubber mallet he had kept hidden, tucked under his arm, until now. Ray watched in horror as he brought it crashing down on the man's hand. There was a horrible crunching sound and a fine spray of blood spattered the floor. The man howled and writhed, desperately trying to pull the arm out from underneath the straps.

Dominic grabbed the man by the hair to hold his head still and extracted the implant.

Then he walked over to Ray.

"Now, this isn't strictly to code. I'd never do this with a made man. But you're just a rookie and possibly still in need of impressing. And I'd like an honest report about its effects from someone I can trust, a guy on my team. So lean over."

Ray gaped at him, not willing to understand.

Dominic rolled his eyes and pushed Ray's head down. Ray felt the click-clip of the implant going into the port echo into his skull and down his spine. He took a deep, shuddering breath

— and his world exploded in pain. He leapt off the chair, screaming, grabbing at his left hand, drawing it close to cradle it, his whole universe shrinking into a solid mass of anguish, falling to his knees, feeling the bone chips grinding together as he moved his hand —

and then… nothing… there was nothing. All of it gone. Meanwhile, the man still bucked and whimpered in his seat, sobbing and shrieking with every pulse beat.

Ray reeled upward and collapsed into his chair, stroking his perfectly intact hand, staring at it, unable to believe Tomasso hadn't somehow smashed his too.

"Oh, yes," Dominic said, blowing his cigar smoke out in a long, satisfied breath. "That will do nicely. Very nicely indeed."

SETH

Seth walked slowly into the tavern, every muscle in his body aching.

It was a typical mediaeval setting. Low ceilings, wooden beams, and dark except for the blazing hearth and a few lanterns scattered throughout. Seth didn't even bother pretending to get an ale, as most players would. At this point, he just wanted to sit down. He found a table tucked away in a corner and sat on the bench next to it. Triggered by his location in the game, his exoskeleton locked into position to support him as though he was really sitting on a hard surface.

He was working on his eighth character as a game farmer, this time, a Mage. He had been looking forward to being one, as every Mage he'd encountered so far was incredibly powerful and vindictive. But it was turning out to be ridiculously hard to level up.

The first time he had spawned in the game as a Mage, he had turned up in a dragon's nest and ended up being baby food. The second, third, and fourth time he had logged in, he'd been killed by a clan that had worked out where a novice spawn tile was. They had parked next to the tile and racked up easy kills. He noted it for the developers to fix.

On his fifth try, he got a new spawn location and

promptly stumbled into a cave full of supersized huntsmen spiders. That game cycle was still giving him nightmares.

On his sixth login, he managed to live long enough to gain his first enchantment. A freezing spell, it was supposed to temporarily halt his foes so he could get away. A great idea, except at his level, it only worked thirty percent of the time. He was getting a bad case of casting arm at this rate.

Seth stared into the flames of the fireplace. He wondered if he wouldn't make more money on a montage video of all the ways he was being snuffed. He was also getting a clear understanding of why upper-level Mages were so mean.

At least it paid well.

"Hail, traveller, well met. May I?" A warrior, dressed all in red, appeared near his table.

Seth suppressed a sigh. He didn't want to talk to anyone right now, but this could be a non-player character with information he'd need for his next quest. He gestured to an empty spot on the bench.

"What, pray tell, troubles thee?" the thief said, sitting down.

Nope, not an NPC. Their dialogue wasn't that cheesy. Seth knew it was against the unwritten rules to break the illusion, but he wasn't sure he could handle more than five minutes of faux-chivalry. "Sorry, just exhausted. My RL name is Seth. What's yours?"

The other character sagged a little. "Mike," he said. "I'm in Toronto. You?"

"Toronto as well," Seth said.

"Wow, lots of people in here from the city," Mike said.

"Well, the game headquarters is here," Seth replied, thinking that should have been obvious.

Mike scratched his belly. "In here avoidin' the father-in-law. My family think I'm working overtime. He's been at our place for two weeks already. Only so many stories about his glory days you can take, you know?"

Seth nodded sympathetically. He didn't have many annoying in-laws, but he had several relatives who would fit the role. They were quiet together for a minute, but Mike was keen to keep the conversation going. Seth could practically see him fishing around for potential topics in his head.

"So how come you're so tired? I always find it kind of relaxing in here."

Seth hesitated. He didn't want to say he was a game farmer. Selena had said it wasn't a secret, but he didn't think he was supposed to advertise was what he was doing. He also didn't want to say he was an author. In Seth's experience, when you mentioned that, few people wanted to ask you about your work, they wanted to talk about something they would write... someday. "Oh, well," he said, "been finding it hard to balance all the things I have going on in real life. And the Mage character is pretty demanding."

"You should try meditating," Mike nodded, looking strangely excited. "Best thing to restore energy. Helps you get clarity."

"Yeah, I've heard about it. Never really got the hang of it. Can only do it for about ten minutes, then I go stir-crazy."

"I used to be the same way. But ... then I got my hands

on something seriously underground. Totally changed my meditation game."

Seth braced himself for a sales pitch. Even though he'd only been playing a short time, it wouldn't be the first time he'd been stuck in a conversation with someone who claimed to be a player but who was here to market the latest service or product.

Mike leaned forward and dropped his voice. It was obvious from his body language, even filtered through the game's avatar, he was bursting to share. "Buddy of mine at work, he let me know about this secret club. Digital Buddhas, they call it. Brand new. They got this thing, see, plug it into your 'jack, and it plays a recording directly into your brain. And at this club, they've got a recording of the meditation state of a real Tibetan monk. You get to use it for ten minutes, and it's supposed to be the best meditation session you've ever had," he said, tapping his temple. He winked. "It's amazing. I've never felt better. But that's just between you and me, eh?"

Seth twisted back and forth to ease his lower back. No sales pitch seemed forthcoming, so he cautiously asked, "It plays a recording of a meditation? And then your brain goes into a meditative state immediately?"

Delighted he'd found someone who was interested, Mike edged closer. "Yeah, it's the darndest thing," almost a whisper, "you sit there, like you would to meditate. They pop this hardware in your jack, flip it on and bam! Instant gamma waves. It's amazing." He settled back in his seat a bit, looking put out. "Wife thinks it's a scam. Doesn't think it's worth the monthly membership fee."

He wasn't so sure 'bam' was something he associated with calming meditation, but the idea of a device like that sent Seth's mind racing. What if he could have recorded the time he was cranking out the words for his book a few weeks ago? Was that even possible? What would it do for his creative output rate? Or his learning rate?

"So, this club, where did they get this device, do you know?" he asked.

"No idea," Mike said. "But you think it would be worth a monthly fee to have access to that kind professional-quality meditation, don't you?"

Mike was looking a bit sulky now. Seth wondered what it was his wife had said to him about the club. He probably would not give him much more helpful advice without validation, so Seth said. "Oh definitely. Say, if a guy was interested in checking this out, where would he go?"

Mike beamed. "New place called DB just off Church. Tell them I sent you. I might get a referral fee for it!"

KEL

"Now in here," the guide said, "we will give you a virtual reality tour of what it would be like to take one of our augmented-reality tours. Specifically, we're going to take you on a tour of Paris, France and show you the view if you downloaded our Nomad pack while you were there." She donned a pair of stylish glasses and pointed at them. "On site, these will beam images directly on to your retina, giving you a heads-up display of information about your surroundings. As you tour the site, it will detect which data points you linger on and customise the display to your preferences. This data is keyed to your location, and also time of day, so it will know the reason you're lingering on information about pain-au-chocolat is because it's breakfast and you're hungry—"

Everyone in the tour group laughed.

"—and it won't drown you with trivia on the history of the patisserie. Unless, of course, you ask it to. If you'll follow me into the arena?"

The tour guide walked backward until she was sure her group was following and then she took them to a large room with several dozen VR goggles lined up in neat rows on the floor. Kel dutifully shuffled in with everyone else.

"Everyone please find a spot and grab some goggles. EduTain is famous for offering different types of AR in the same package, to accommodate different learners. So if you're an audio kind of guy who hates reading, you can toggle to that mode. Okay, I'm going to join you in a minute and put mine on. Everyone got theirs on? Everyone seeing Paris right now? Good. And, as I was saying with respect to learning, if you're a people person, you can toggle and you can interact with people on the tour. So if history is your thing, you can chat with Clovis the Lazy here about how things were circa the year 632."

An image of a young-looking Frankish king appeared on the modern-day Rue des Barres. She pulled off her goggles and glanced around. Everyone was lost in VR, staring into their viewers and exclaiming at what they were seeing. The guide was also immersed in the imagery.

She quietly put the goggles back on the floor, threaded her way through the crowd, and slipped out of the arena. The official tour wasn't telling her anything she didn't already know about EduTain from a few searches. She had to figure out what they were working on, what they were researching. Why were they tracking her lab? Had they stolen her device?

A security guard looked at her closely as she crossed the main hallway of the building, so she made a beeline for the washrooms she had seen there earlier. Once she was safely behind the door, she pulled it open a crack to look out. The guard had lost interest for the time being and gone back to watching the doors.

Kel removed her visitor badge and tucked it under a

towel on the sink. She'd be less conspicuous without the bright yellow label.

She left the washroom, ducked into the hallway where the lifts were located. The tour guide had said they did all of their research and development on the sixth floor, so it was the number six button she selected when a lift arrived.

The lift doors opened onto a quiet but busy floor. There were a series of large offices and workspaces, all with huge picture windows that looked into the hallway, and even larger windows that provided a view of the outdoors. Given the amount of activity she could see going on, and the lack of noise, Kel assumed they were well soundproofed. She was immediately jealous; she much preferred this arrangement to the old-fashioned open plan layout of her own workspace.

Kel strolled down the hallway, trying to look as though she had a destination, but stealing glances at rooms on both sides. One room appeared to be used for music scoring and sound effects. In another space, several people were clustered around a wall, arranging a storyboard. In still another, costume designers were busy testing period costumes to be scanned and rendered into virtual reality.

Nothing looked promising until she rounded a corner and discovered the space where they were working on VR hardware. The workshop was full of parts, fabbers, and prototypes for next-generation eyewear, controllers, and feedback systems. Kel paused, looking at the various bits and pieces spread over a workbench. Was her device there? What would they want to use it for? There were just two people in the room right now. Could she hide in the building until later and come back to this area when they had gone for the day? She looked for storage rooms.

"You there!"

Kel spun around. It was a very large security guard. "Uh, hi," she stammered. "Which way to the washrooms?"

"You know that already," the guard said. "Come with me, please."

"I do?"

"Yes, you left your badge in the washroom. The system is heart rate synced. We knew the minute you dumped it. Come with me. Now."

~

If she had been jealous before, Kel was green with envy now. This woman's office was enormous.

"Thank you, Manuel. You were right to bring her to me," the woman said. "I'll be sure to update Ms Torres on this when she gets back. Could you wait outside?" The guard nodded, then left Kel and the woman alone.

"Please sit down," the woman said, pointing to a chair. "I'm Pauline McDonald, executive assistant to Maura Torres." Kel sat down. "So which organisation are you spying for?" Pauline demanded.

"Me?" Kel replied indignantly. "Why is *your* company monitoring our computers?"

Pauline cocked her head to one side. "What organisation are you with?"

"You mean you're bugging so many you don't know which one to start with?"

Pauline didn't answer, and the two women glared at each other.

Kel finally said: "I'm Dr Kel Rafferty, University of Toronto. I'm doing Alzheimer's research. My... something I was working on has gone missing. I think you have it."

Kel thought Pauline looked surprised, though the expression was so fleeting it was hard to tell. "What was it you were working on? What could possibly be of interest to us?"

"I would rather not say, and I really don't have a clue. All I know is that I found your packet sniffer in our network."

"I have no knowledge of any packet sniffers," Pauline said.

Kel thought that was probably true. Someone who worked as an assistant to the company's CEO likely wasn't involved in setting such a thing up. That didn't mean the company hadn't deployed any, though.

"While I'm sure the work you do is interesting and noble," Pauline continued, "I can't imagine why you'd think we would be interested." She leaned back. "And if you are missing a device, I suggest you take it up with the police or make enquiries on the street. But please don't go wandering through private property, Dr Rafferty. We take corporate espionage quite seriously. If we find you poking around here again, we *will* have you arrested. Manuel will see you out."

SETH

Mike's secret club turned out to be a drab storefront on Church and Gerrard, right beside a clothing shop that featured several gynoids and androids modelling the latest fashions in the window. The proprietor hastily wrapped himself up in a dingy saffron robe when Seth came in.

"Namaste," he said. "How may I bring peace to you today?"

"Hello, is this Digital Buddhas?" Seth asked him.

"Indeed, it is," the man said. With blond hair and blue eyes, he didn't look very Tibetan to Seth. Then again, he'd been told he didn't look Italian, so who was he to judge?

"I heard about this place from someone I met in VR," Seth said. "Told me you had this thing you could plug into a standard jack for a fantastic meditation session."

"Ah yes, we have several packages available. Our Instant Transcendence package is our best seller. For the price of just one great meal out with drinks, you can—"

Seth raised a hand. "Actually, it's not transcendence I'm interested in, it's—"

"Of course, if that doesn't suit your budget, you can try the next level down, which is our Chakra Charger. You will feel better, attract more—"

"No, no," Seth said, "you don't understand. I—"

"For bargain hunters," the man said pointedly, "we have a Healing Hertz for positive thoughts. This is being offered at the introductory price of—"

"The device," Seth interjected. "I'm interested in the device. Are you selling that?"

The man bristled. "Are you suggesting you wouldn't get a genuine encounter here?"

"What? No," Seth said. "I don't care about the experience. It's probably great. I just want to know more about how it's done."

"You doubt my credentials, is that it? Going to start up your own shop? Where did you study, eh? I spent three weeks in Tibet!"

"Good grief, I'm not interested in anything to do with meditation. I want to know where you got the device."

The man pointed to the door. "Out!"

"Jeeze, you're not very zen," Seth said.

"Out!"

He decided to pop into a few of the other shops and ask questions. News of the device had spread like wildfire down the rest of the flow, but no one here seemed to know where to get one.

He decided to try one last place and stepped into a place called The Black Eagle. It was a Saturday, so the place was busy with prospective customers. He was accosted by a rentbot set to aggressive sales mode. "Hey, sailor!" he said. "What are you into tonight? Did you know our loyalty club gives you a permanent ten percent discount? What do you say?" He twirled around, showing off a perfectly proportioned swimmer's physique. He stopped and posed seductively. "I'm available right this very minute."

"Sorry, not here for a rentbot, thanks," said Seth, and the bot pouted. "But I am looking for information on a special kind of recording device."

The bot rolled its eyes. "Everyone has been asking about those in here today. I'm beginning to think I will be out of a job soon. Follow me."

Seth followed the bot to a display shelf at the side of the showroom. There were several objects locked behind glass. The bot undid the cabinet and pulled one out. "We're giving these away free for a limited time. Your mission, darling, should you choose to accept it," the bot said, with a charming smile, "is to record your hottest experience, start to finish. Bring it in and we'll give you three months' exclusive access to our Love Exchange club for free. And you'll be enrolled in our loyalty club at no charge."

"And what if I don't bring in an… experience? What happens to the device? Does it expire or something?"

"Sweetheart, don't sell yourself short! You have plenty of time to find a hot date. And remember, I'm not booked yet!"

"No seriously," Seth said. "What happens if I don't?"

The bot let out an exasperated sigh. "The boss claims there's a ninety-day lockdown installed on it, so you must bring it back in then. But honestly, according to my scans there doesn't seem to be anything like that on there. I am certain he hasn't the slightest idea how they work. He's just fabbing copies of the device he got off someone else. Just recycle it when you're done. Now if you'll excuse me, I'm off to find someone much more willing."

Seth examined the device. It looked innocuous and like it would fit into a standard port. He'd take it home and let Tasha examine it for viruses and other malware before trying it out.

HAROON

At another recruiting event, Haroon hung around until the officer wrapped up before he came into the room. He approached the woman nervously and waited while several people from the audience asked her questions.

When the last person finally left and he got his turn, she smiled warmly at him, which gave him the courage to speak up.

"Hi," he said, "they told me to meet with someone here about my application?"

"Ah, right," she said. "I'm Constable Martin. And you must be…" She picked up her tablet and tapped it a few times. "You must be Haroon. Thank you for coming out to meet me in person. We got your application and your medical exams. I wanted to communicate with you personally. Your grades are decent, and we really liked your enthusiasm in your application. Your medical exams checked out, but, well, I wanted to discuss your police record with you."

Haroon was bemused. "My police record? What do you mean?"

"Typically, we don't look at someone with a history of serious criminal offences, including involvement with organised crime, but given you were underage we were wondering if there were any extenuating circumstances that aren't stated in the reports?"

Haroon opened his mouth and shut it again. It was several long moments before he could process what she'd said to him.

"But I don't have a record!"

The constable's expression was confused. She looked at the tablet and then back at him. "Are you trying to claim you are not Haroon Subhan Minhas?" She handed him the tablet.

Haroon viewed the report and nearly did a double take. His own image looked back at him.

Except, when he studied it closely, it wasn't him at all, but rather, someone who looked very like him, twenty-one years ago.

"That's my father," he breathed.

"I'm sorry? Could you repeat that?"

He swiped the report up to read it. They had caught Subhan in a police sweep just over two decades ago, not long after he had moved into J-District. After a brief scuffle with the arresting officers, and a bunch of lengthy interviews, he had admitted to working as a part-time security guard for the mafia. They released him on bail and then he was rearrested several days later for public drunkenness and assault. The second arrest report suggested he had many severe injuries that were a few days old, and some new ones from the fight he'd been picked up over.

Haroon abruptly recalled one of the many times his father had woken up screaming over the years. "I say no thing!" he would shout. "I tell them no thing!" He must have been worked over by the local don to find out if he'd said anything incriminating to the police. He eventually spent time in jail for his association with the gang, but they made no other major arrests connected to the case. The mob apparently decided he hadn't snitched after all.

"Mr Minhas," the constable said. "Could you say that again?"

"This," Haroon said, still reading, "this isn't me. This is…" he paused. He wasn't willing to admit this was his father a second time. "Look at the report dates here. I'm only nineteen."

The constable took back the tablet and studied the report. "Hmm, look at that. When records with similar or identical names are pulled up, the computer usually filters by image to reduce false positives." She glanced at his photo and then at him, back and forth, even holding the tablet up near his face. "So why is this your picture?"

"It's not. Look, the nose and chin are different," he said. She didn't look convinced.

"Let me see your ID please?" she pulled a small object out of her pocket and scanned it over his wrist. She referred to her tablet again. "You didn't get this ID until just a few years ago? Forgive me, but I'm wondering why that would be? An ID with an age so far off chronological age is usually a sign of ID theft or something worse."

Haroon could sense a shift in attitude, going from friendly recruiter to someone investigating a possible crime. Her tone wasn't hostile, but it had gotten a lot sterner in the last few minutes. "I wasn't given one at birth," he replied. "I never knew my mother, and grew up more or less on my own." It wasn't a lie, but it glossed over his father. "If you check dates for my education, you'll see I had a late start on that, too."

"Oh?" she said. "Yes, it lines up. With the same name and what looked to be the same face though…" She squinted at the image and then shook her head. "A relation perhaps?"

Haroon shrugged. "Maybe. It's a big city. I suppose I might have family around, but…" he let it trail off.

She gave him a shrewd look. Haroon wondered if she guessed what the relationship might be. "Well, the force would reserve the right to ask for a test to prove your age should you go through with your application and on to the next step which is the entrance exam. In the meantime, I'll disassociate this record from your application. Now I'm really glad we took the time to talk to you in person. I would hate to lose someone because of some confusion over dates, and the way you handled yourself now suggests you would be a good candidate. Not everyone would have taken being confronted with a record — even an erroneous one — nearly so well. Being calm under pressure is a key part of the job."

"Thank you," Haroon said.

"May I send you some test dates?"

The last thing he felt like thinking about just now was an exam, but he didn't want to say as much to her. "Yes, that would be fine."

She forwarded the information to his wristband, they bowed politely to each other, and Haroon left the building.

He wilted onto a nearby park bench, ignoring the rain soaking through his trousers.

His father was a criminal. It looked like he had chosen — chosen! — to move into J-District. Haroon had always assumed Subhan had grown up there and hadn't found the means to escape it.

Ever since he was old enough to understand the term, Haroon had known his father was an alcoholic. A part of him had always worried he would be one, too, which was why he'd never been tempted to sneak a drink. He knew there were good treatments for alcoholism available — *outside* of J-District — so if he got out of J but in trouble with booze he still had options. Yet still, for many years in the small hours of the night, he had wondered if he could leave the district, and even whether that treatment would be enough. What was it was like, not being able to say no to a drink? Would he ever feel like he couldn't?

This, though, this felt different from alcoholism. It wasn't an affliction or a disease, or even stealing food to feed your kid. This was working with the mafia. The same people that once threatened Yoshi and terrorised everyone else in the neighbourhood for years.

His own father.

And what about his mother? The one time he'd asked his father about her, he'd been backhanded across the room. What happened to her? Was she a criminal, too? An alcoholic? A drug addict? Was this why he struggled so much to concentrate in school? Why he always felt so fidgety? He wasn't stupid. He knew the fact she wasn't even in their life or nearby wasn't good news. What if her record turned up suddenly? Haroon thought of the recruiter's increasing brusqueness. What would the police think of him then?

Was there any part of him made of anything good?

MAURA

"No, no, no!" Maura's fist crashed into her desk. Then, embarrassed by her loss of control, she took a few deep breaths.

Pauline had just come in. She changed course and went directly to the fabber for tea. Maura accepted it gratefully.

Pauline sat down in the chair by the desk, leaning forward. "So what can I do?"

Maura gave Pauline an irritated look. "Help me turn back the clock? To prevent this?"

If Pauline was affronted by the snark, she said nothing. She simply waited until Maura was ready to discuss the problem.

Maura sighed. "Xperience has just declared bankruptcy and folded. No one knows where the CEO is — he's probably skipped the country or something. He will have done if he's smart." She sipped her tea. "Then again, if he was smart, he wouldn't have driven the company into the ground."

"Talk me through what this means for us?"

Maura sat her tea down a little too sharply, and the cup rattled in its saucer. She shouldn't take it out on Pauline, she knew. Maura had dug, and dug again, and still couldn't come up with anything that hinted Pauline might be a spy for another company. And yet… she couldn't shake the nagging feeling her problems had started when Pauline had shown up. First, the discovery of that dratted implant, and now this.

Maura studied Pauline, wondering what to say. She was a good assistant. Too good, in fact. Pauline had developed a way to help Maura articulate things more clearly and come to decisions more swiftly. She also anticipated Maura's needs in a way previous assistants hadn't done. She was getting rather fond of the woman, and that bothered her. She didn't want to become close to anyone.

Why was Pauline here, really?

"For a start," Maura said, trying to focus on the present, "it means our attempt to take over the company dies. The assets will be tied up for ages in bankruptcy court at the very least, unless their board of directors comes up with a plan for the creditors. And if there is a process worked out, it's unlikely to be something that directly involves us. Any creditor with a claim on the company will get what they can for a fraction of its value and then, assuming it's not their line of business, put what they get up for auction. If we are successful at auction, it would be because we will have paid through the nose for something we could have gotten through acquisition much more cheaply." She sighed, aware that she sounded like she was ranting. "To say nothing about now not being able to acquire the structure and staff of the company in one go. And if we weren't successful, it would be because our competitors nabbed assets we wanted."

"Okay," Pauline said. "I can talk to HR about headhunting the best of their staff right now. But they weren't our only acquisition target."

"No, I suppose they weren't."

They were silent together. Then Pauline said, "So what else is bothering you so much about this?"

Maura got up to go to the window to look out. She felt so restless. What *was* bugging her? This was business after all. Running a company was like riding a roller coaster. There were ups and downs all the time, sometimes terrifying ones. This was a setback, but nothing more, for EduTain. Wasn't it? There had been setbacks before, some worse than this.

"I suppose it's that all of those people are out of work now. And the Dragon Slayer game was a good product that gave customers joy. I hate to see potential just destroyed, for no reason. Well, for any reason. But especially when it happens because someone was an idiot."

Pauline looked thoughtful. "That's a good explanation, I guess. But lots of people lose their jobs for stupid reasons all the time. I'm curious why it bugs you so much. I thought you'd always had this company? Did you lose a job once before that, in similar circumstances?"

Maura turned and regarded the sculpture on her desk. "It's true I've never personally lost a job. I inherited EduTain from my father. It was a complete mess when I took it over."

"Ah," said Pauline. "I'm guessing people lost their jobs at some point?"

"No," Maura said harshly. She wondered why she was

getting into this with Pauline. She'd never discussed it with other assistants. Yet she couldn't stop the words from tumbling out. "Yes. We lived in Columbia. My father had a good heart but no head for business. And no aptitude for politics or keeping an eye on the political situation or…" Maura drew a ragged breath. "He was a bon vivant, as they say. Existed in the now. Our house was a constant stream of visitors and dinners and parties, or dashing off to visit other people at the drop of a hat. We took long trips almost weekly. I'd never know where I'd be sleeping night to night. Which probably sounds exciting, but…" Maura stopped.

"What happened?" Pauline asked quietly.

"It was Columbia. People forgot what it used to be like back in the twentieth century, with the various guerrilla groups and paramilitaries. When prosperity came to Columbia, people were caught up in their daily, very comfortable lives and forgot about that still very significant group of people that were stuck. The ones that couldn't or wouldn't change their situation for whatever reason. They thought… well, *we* thought, that we could relax, that we were past that, that we could stop being vigilant about the agitators and populists. We forgot to hold the line on human rights. We weren't past any of that. So there were a lot of discontents, ripe for radicalization." Maura squared her shoulders, set her jaw. "So when it all came crumbling down again, my father and mother were killed in the crossfire of a paramilitary assault on a police station. Right in front of me. I was twelve. My mother was eight months pregnant at the time."

"I'm so sorry," Pauline said, blinking rapidly. "I didn't know."

"It's not something I advertise or put in my public bio. That's just not the story I wanted for myself here. I inherited control of my father's company, but a cousin ran it until I came of age. I beat it out of Colombia as fast as I could and took as many people as wanted to go. Canada was the obvious choice. There are so many supports here. I still see people complain about how easy each new generation of immigrants have it compared to the last, or how it somehow privileges newcomers over citizens born here, but that's just comfy people grumbling. Unlike many other places, it is a country that actually wants you to succeed."

Pauline hesitated before she asked, "So you're angry with your cousin for making it a mess by the time you got it?"

"I adored my father," Maura said. "But I also hated him. For not paying enough attention to the brewing trouble around him. For not being in control. The company struggled for years afterward and many people lost their jobs. For not making provisions."

For getting killed, she thought. For leaving me.

Maura shook her head, now furious with herself for saying so much. What had gotten into her? She sat back down at her desk and finished her tea. "And now you know," she said tersely. "So let's get on with the business of dealing with the situation we've been dealt, shall we?"

MEIKE

Meike sank into the water, the liquid purling over and around her shoulders. She enjoyed the way the heat and alcohol were making her feel dizzy, on the verge of passing out. Her head lolled.

Lorenzo, sitting on the side of his hot tub, sipping champagne, kicked at her hip. "Don't die in my tub. My cousin hates body disposal jobs."

Grudgingly, she sat up and shifted sideways onto a higher seat so her torso was above the water. The palatial bathroom felt cold by comparison. She drank some more champagne.

Lorenzo hauled himself out of the tub, the water sliding off his hairy body and puddling on the tile floor beneath his feet. He grabbed the champagne bottle and padded over to refill her glass. He was a big man, with exceptionally broad shoulders, a narrow waist, and strong legs. Lorenzo wore his black hair wavy and longish, and his eyes were dark and aloof. He had a three-day growth of beard encircling an expression that was almost always a slight frown.

Meike took the refill and drained her glass, holding it out for another before he could walk away. "*Impudente*. You're buying the next bottle. You can afford it now."

"Yes, but you can afford it more, and mine won't last," she said. He smirked.

They were in Lorenzo's luxurious apartment in Lawrence Park North, an area she had only ever driven past before now. A well-connected man, he had friends in both high and low places, because he liked slumming even more than he liked living it up. They had met at the appropriately named Club Débaucher; she had arrived with Fa and left with Lorenzo. And that was that.

He got back into the tub and regarded her with half-closed eyes. "You want to afford it more?"

She shrugged. "It would be nice. But I don't think he'll give me any more than what I got."

"Tomasso says his boss is very pleased. Finding multiple uses for it. You could probably have asked for more." He belched. "But that's penny-ante stuff. I'm talking big money."

"I'm not going to deal, if that's what you mean. I'd be my own best customer."

"Nah, hardly anyone bothers with anymore. Pocket change." He swigged straight from the bottle. "You've got this look though, like nothing I've seen before. It's… what's the word? Androgynous. Malleable. It's both sexy and weird."

"Thanks," she said sarcastically.

"I mean it. You could be anyone."

"It's a good thing you didn't use this line on me in the bar, or you'd be going solo right about now."

He laughed, a short, humourless bark. "What I mean is

you'd be great in the holos. You could play any character you liked. You're a chameleon."

Meike closed her eyes and thought about it. She'd watched a few holos, like anyone else, and been amused by some. It was sometimes fun to escape whatever was going on in the real world into an alternate universe for a while. She had never considered acting in one.

"Aren't most of the characters in those things generated?" she said, still keeping her eyes closed, the mushy-headed feeling she had now verging on queasiness.

"The low end stuff they churn out for the formula shows, sure. They've got stock characters and a big enough variety of interactions they can just about make those work, especially because the audiences aren't too picky. But the features and so on? All human actors. They bring the creativity and unpredictability to the performance."

It did sound sort of interesting. Being Meike wasn't all that engrossing most days. Pretending to be other people on a regular basis had a certain appeal.

"How would I get into it?" Meike asked. "I can't act." Acting also sounded like an awful lot of work.

"I know a guy," Lorenzo answered. "I'll get you hooked up. And you won't have to act, really. You'll just have to channel whatever we can plug into the back of your head."

Meike thought about it for a moment. "Oh, I see. Clever. And what do you want out of this?"

"Finder's fee, a permanent percentage," Lorenzo said, honestly. "But I'm in it for the long game. I'm looking at access to a celebrity. I want into that crowd."

Now it was Meike's turn to laugh. "I'm not anybody yet."

Lorenzo took a few big gulps of champagne, spilling some down his chest. He wiped his mouth with the back of his hand. "Pretty sure you will be."

RAY

Ray had spent three frustrating weeks on gambling-den duty, which had involved little more than fetching expensive drinks and snacks to keep the high rollers at the table and losing money way into the wee hours of the night. It was a chump's job, one that kept him exhausted and footsore.

In his downtime, he pestered — as much as he dared — the other guys in the organisation to teach him how to hack. He spent hours learning and practicing. On the surface, he looked like any other new guy, aiming to skill up to please the boss. In reality, he was trying to find out about Mick.

Getting into the police database turned out to be shockingly easy. An old networked coffee machine in a local office hadn't had a firmware update in years and thus was wide open. He ran dozens of queries using a random date generator to mask his true intent, but only read the entries for the day of the bombing.

The file he wanted was small and unhelpful. The drone in the explosion had been part of an entire fleet that had been decommissioned in favour of newer models and slated for recycling. That device had been remotely hijacked from a pile somewhere, filled with the explosives and shrapnel, memory-wiped to that point, and launched from behind an empty warehouse along the lakefront.

Of course, there was nothing in the file on Mick or Ray, apart from their victim status. They were blanks. Not even the incident with the overzealous cop in the hospital, not that Ray had truly believed that would have been detailed for posterity.

He ran searches for information about the businesses in the area. Where had Mick been heading? One tantalising lead cropped up: the highjacked drone had once belonged to a company downtown, close to where he'd seen Mick that day. Had Mick gotten mixed up in a bad business deal? Ray wrote it all down slowly, painstakingly into his notebook; his untutored printing like that of a child's.

Then Ray had received another gut-chilling summons from Dominic.

Seated at his desk, his feet up, Dominic sucked on his ever-present cigar. "In three weeks, I haven't heard a whisper about our last meeting, my little experiment. That's quite the secret to keep, this magic device. I knew I could trust Tomasso; now I know I can trust you." He picked up a velveteen bag from his desk and tossed it at Ray. "Inside: six implants for replay. Each has a ten-second delay. And keys to your new apartment. You'll find it much improved over your last one. Be moved by Friday."

Ray peered inside the bag.

"I've got ambitions. I've got a small empire now. I want a massive empire. I want a reputation. You are now my *mago*, Raymond," Dominic said. "That room is your domain. We will be bringing in lots of guests. We'll impress upon them I am a person to be reckoned with now. I can bring the pain, in a way no one has thought possible before. And it must involve theatre. Dress in black. Shades for your eyes. Say nothing. Just insert the implant when Tomasso gives you the signal. We start today. Now go."

Ray nodded, grateful to the point of being giddy not to be the subject of any more experiments, or worse, he hustled back to his current apartment and changed. He returned to the plaza and waited in what he now thought of as The Room.

The waiting gave him plenty of time to think. This new assignment was a relief. It would keep him near the centre of all the decision-making and give him a chance to eavesdrop and learn more about the city's underworld. He was also happy to know he wouldn't be put out on any more missions like his initiation. Separating gamblers from their money wasn't so bad, as he figured they were there by choice and knew the risks. But stealing by syphoning like that? At that scale? He'd grown up stealing to survive, but this was... different somehow. And this new job didn't seem like it would be too bad either. It wasn't like he was beating someone up. He wasn't sure if he could hit someone who wasn't trying to kill him.

There was only one person he truly wanted to hurt, and that was Mick's killer. Afterward, he planned to disappear. He hadn't worked out how yet, but he wanted to be gone. Out of J, maybe even out of Toronto altogether. For once, being a nobody might help.

After about an hour, Tomasso brought in a heavyset man

with tattoos across his knuckles, cropped, greying hair, and an attitude. Tomasso, one hand inside a pocket, shoved him roughly down into the chair. The man spat at Tomasso's feet. Tomasso just sneered at him. He looked at Ray and snapped his fingers once. The noise echoed in the big, empty room.

Ray thought he had worked out what they expected of him. In his brief encounters with Dominic, he had noticed one thing: the man liked style. Ray stayed silent and walked deliberately from the side of the room, first walking towards the man who was facing him, and then slowly, ever so slowly, around behind him. He clicked the implant into the port, trying not to shudder when he remembered how it felt. He timed his walk back around to face the man again to be about ten seconds.

No sooner had he stopped, the man jerked, clutching at his stomach, and gasped. He looked at his hands in wide-eyed horror. Then he shrieked, and his hands clutched again, this time scrabbling across his gut, following the pain. He slid down out of his chair, jerking and twisting.

Ray realised the man thought he had been stabbed.

Now completely out of his mind, the man wailed, squirming, his hands now desperately working as though … as though he was holding his intestines in.

Ray's mouth filled with the remains of his lunch. A sharp-eyed glance from Tomasso made him swallow rapidly, quivering.

The recording lasted several more seconds. The man was sobbing, feeling, but not seeing what he was feeling, all trace of his former toughness crushed by confusion and gibbering fear. Ray's mind kept racing back to the pain of a smashed hand and his fist clenched and unclenched.

At last, the replay finished. The man collapsed, his face drained of colour. Then he rolled over and threw up, bile and vomit spattering the floor. Behind his glasses, Ray closed his eyes and willed himself not to add his own mess.

Tomasso waited, indifferent, until the man had finished retching. Then he crouched down, casually putting one hand on the man's neck to withdraw the implant and whispering something into the man's ear. Sobbing, the man nodded vigorously.

Tomasso made him stand. The man turned and caught sight of Ray and stepped back abruptly, crossing himself. He waved a quivering finger at Ray as Tomasso shoved him out the door. As they left, Tomasso turned and made an imperious gesture at the mess on the floor.

It took Ray half an hour to clean up because he kept heaving and choking. It smelled awful. He had only just finished when the door opened again.

It was Tomasso.

And this time the person with him was a young girl.

KEL

Kel wasn't sure how one went about making enquiries on the street; it was such an old-fashioned term, and it sounded like something out of a bad holo. But given her only other option was going back to EduTain and risking arrest, she didn't think she had any choice but to figure it out.

When she got away from her work, she returned to the Bitz shop where she had her port installed and asked some vague questions about new implants for standard ports.

"Getting bored with your toy already?" the woman who'd spoken to her before grinned. "We haven't got in any new stock, but I had lunch with the guy from the piercings place across the flow yesterday and I think he mentioned something about a new thing he was looking forward to trying. He might have the dirt on the latest and greatest."

Kel left feeling happy. If the top port-install place in Toronto hadn't heard of what she'd created, it wasn't loose in the wild yet. She still had time to figure out who had stolen it and find some way to get it back.

Back outside the shop, she took a moment to inhale the cool air. It smelled like spring, fresh and new, and full of promise. The sensation, as always, made Kel feel a sense of urgency about her work, and how much she still wanted to do. Time was passing. She considered going home but then figured she should go talk to the piercings shop owner. Kel knew if she didn't, she'd be awake at 3:00 a.m. wondering if she should have double-checked. Better to be thorough now.

She walked over to the crosswalk and waded into the traffic, and the pods formed a bubble space around her as they stopped or slowed to let her through. When she stepped into the piercings shop, she was halted by an overpowering wave of sandalwood incense. A man with six facial studs arranged in a curve down his cheek, starting with a large one and getting progressively smaller, greeted her.

"Hello," she said. "Your friend across the flow said you might have a new item you were keen to try."

He raised one eyebrow, making a couple of his studs surge upward, too. "What, like a new set of rings or something?"

"No, no," Kel said, pointing at her port. "A new implant."

"Oh, that," he said. "Not sure why she'd tell you." He went behind the reception counter and opened a drawer that rattled with lots of various bits and pieces. "It's not something I'm selling. I'm just going to experiment with it here." He pulled out a small black device and put it on the desk.

Kel nearly gasped aloud. She could see the intricate port interface. It was printed with a different material for the case, but it was unmistakably the brain state recording implant she designed.

"Where did you get this?" she asked, her voice barely audible.

"My mate at ReLeaf gave me one," he said. "It's supposed

to record what your brain is doing. I'm going to try it as a substitute for anaesthetic for some of my more squeamish and sober customers."

"No!" Kel almost shouted. The man looked at her with a combination of alarm and suspicion now. "No, you mustn't. It's not tested for that."

"How would you know?"

"Because…" Kel paused. There was nothing for it. "Because I designed it. Could you give me it please?"

"Why would I do that?"

"Because it's mine and it shouldn't be out yet. It's not ready."

He crossed his arms. "Some weirdo comes in off the path claiming she designed something, and I'm just supposed to hand it over? I don't think so."

Kel didn't stop to think. She dashed forward, snatched it off the counter, and bolted out the door. She was around the corner and out of sight in minutes. She overheard the man shouting abuse some distance behind her.

Once she was sure she wasn't being pursued, she stashed the device in her bag and used her wristband to find out what and where ReLeaf was. She had to move fast, before the first man got in touch with his friend, warning him, describing her. She toggled her jacket from its current bright red colour to a nondescript beige. Kel cut across the flow mid-stream, ignoring the screeching of tyres and brakes that came from not indicating her desire to cross first, and made her way down the block. She worked to slow her breathing to something more normal and pulled her hair back into a bun for good measure. It wouldn't stop a serious effort to find her, but at a glance, she was now no longer a longhaired woman in a red jacket.

ReLeaf was apparently a marijuana dispensary. There were only four people in the store, well into enjoying the house speciality, seated in lounge chairs around a table. The air was thick with the smell of it.

"Well hello, darlin'," said one. "What can I do for you today?" He sported an old cowboy hat, had kind eyes and a bushy moustache. He reminded Kel of a celebrity from decades ago, back when holos were called movies and still flat, but she couldn't think who just then. She put on her best smile.

"Hi," she said brightly, stealing a glance at his wristband. It was already flashing to show a message received. Was it the guy she had just stolen from? Would be going to Bitz to complain and get her name from the girl behind the counter? "A friend of yours said you gave him a device for recording the brain. I'm really interested and wondered if I could see it."

He took a slow drag and closed his eyes, savouring it. It seemed an eternity before he opened them again, and when he did, he struggled to focus. "Oh yeah, yeah, I have one of those in right now." His gaze wandered away and his face split into a slow, wide grin.

Kel waited for as long as she dared, the light on his wrist still flashing. "Could I see it?"

He looked over at her, surprised she was there. He waved the smoke away to see her better. "I'm a businessman you know. Smart one, too."

"I can pay..."

But he wasn't listening. "Saw one of these jobbies and thought hey ho, you know? That could seriously change the trade. Smokeless highs. Much better for you." He reached up to scratch behind one ear. "Been trying to record the perfect high since ... since ... when did we start this, Joey?" He nudged one of the other men with a foot, but the other guy was even more zoned out and snoring slightly. The shop owner giggled and looked at her as if to say, 'what are you going to do?'

Kel could feel the smoke getting to her, too. Whatever it was they were enjoying, it was potent. She was desperate to get the implant and get out but she was unlikely to wrest it out of him, even if he was stoned. "That sounds really smart, really blast," she ventured. "Can I see this device? I'd love to see yours. Could you show me?"

"Whadya want to see mine for? You can go get your own. You could p'rolly get three."

Kel stiffened in shock. "What are you saying?"

"Buddy at the Eagle gave me this one. He's even smarter than me, if you can believe it," he puffed. "Got this promo goin' on gonna capture the brain porn market straight up, he said. Free devices in exchange for bringin' in recordin's."

Kel staggered out of the shop into the fresh air, close to tears. In less than an hour, she'd found three places that had her implant, and one of them was handing them out like candy. Someone must have taken one apart and scanned it well enough to create a reliable fabber schematic of it already.

Another thought crashed into her. Pauline had told her to go look for her device.

Kel hadn't said what had been stolen from the lab.

That meant her company probably had her original.

The genie was well and truly out of the bottle. Her only hope now was that none of the people she had just talked to would know who she was and connect her to the mayhem she was sure was to come.

SETH

Seth flicked the implant again, and felt the surge, the glorious surge.

He sat in his new rig, bought on credit, in an immense, synthleather cockpit chair moulded to his body. The stylized back of the chair arced up and over him, looking like a scorpion's tail, its stinger the hook for a rack of ultrathin screens, each displaying a different task in progress.

One was scanning and reading out his favourite poems, the elegiac lines pumping directly into his ears at double speed, priming his brain for lyricism. Another was cranking through an enormous database of reader reviews, analysing, collating, picking out key phrases, learning why each book was adored or hated. What made the reader tick? A third screen he set to work running a plot generator on his own nascent work, inducing possible narrative directions based on what he'd written already. Into a fourth, he poured a corpus of literary works from around the world into an AI, mashed them, mixed them, and spun up dozens of random snippets from the slurry. Some were nonsense, others made him laugh; a rare few sparked an idea that burned so hot he worked feverishly to write them down before they evaporated.

He pulsed through several more replay cycles like this,

each time enjoying the sensation of fierce concentration as it washed over him again and again. Seth paused only once to fab some stimulants and gulp some water. He did not want to let this go, not yet.

He felt like he was glowing, the brightness of his spirit at last illuminating the dim recesses of his mind, bringing forth his best self into the light.

Time passed; he didn't know how much, he didn't care. He folded his rig, donned a new pair of goggles, and watched his apartment shimmer away into a VR environment. Seth ground through a programming sequence almost effortlessly, feeding the elements of his novel into the device, and saw his work extrapolated into life in his living room: as yet crude, primitive, but there, the ghosts in his head made real.

He talked to his protagonist, listened to his villain, and walked through his world. Seth tweaked, sculpted, and choreographed. He made them all interview him, probing his subconscious, astonishing himself by finding answers he didn't know he had.

Then he dumped the sessions into transcript, projected the text on to his virtual walls, grabbing words, sentences, paragraphs with his hands, moving them, shaping them, saving gems, throwing away tailings.

He was a conductor and this was his symphony. Keeping time with the steady flick, flick, flick to reset his implant.

And then at some point, he swayed, staggered, and crumpled into a heap on the floor, breathing hard.

A voice, incessant. He struggled to pinpoint it.

"Seth? Seth?" It was Tasha's voice.

"Seth? I am detecting abnormal levels of cortisol, DHEA, adrenaline, noradrenaline, dopamine, and aldosterone. You have had an abnormal heart rate for three consecutive hours, and your sleep log indicates little sleep overall for the last several days, and very little REM sleep in particular."

He could hear her. He knew she was saying something important. But he couldn't work up enough energy to puzzle it out. His arms and legs felt leaden.

"Your food and drink consumption is down, and you have not queried me for information at your usual rate," Tasha said.

It was silent in the apartment for a while.

"Seth, please respond," Tasha said.

Seth curled into a ball. He was so tired. He didn't want to sleep. Everything felt jagged, jittery and strange things floated in his field of vision.

"Seth, as per my manufacturer's presets, I'm initiating Mental Health Protocol I," Tasha said.

After a few minutes, an enticing aroma reached his nose. He could smell his favourite tea. It was quite a while before he found enough willpower to go to his kitchen. He found a cup and filled it with the liquid in the fabber. He couldn't remember having asked Tasha for it.

It tasted very good. In fact, it was the best cup of tea he had enjoyed for a long time. He drank it all and had another. Suddenly, he felt all loose-limbed and wobbly. Pleasantly sleepy. He only just made it to his bed in time. He fell into a dreamless slumber.

~

When he awoke, he was famished. Tasha printed him a big breakfast without him even having to ask. It was delicious.

It wasn't until he'd checked the news that he realised he'd slept for almost twenty-four hours. He checked Tasha's logs. She'd put a mild sedative in his tea to help him sleep, and this morning's meal had been supplemented to help prevent symptoms of depression. Seth realised she had music playing — a cheerful, but not overbearing instrumental piece.

After a hot shower, he sat back at his workstation and stared, goggle-eyed at what he'd done. There, right there, was three-quarters of his novel, complete, and dozens of ideas for additional books. It was all still rough, to be sure, but it was more progress in the space of several days than he'd made for months. Seth checked the logs, not trusting his memory. Yes! There, in the VR sessions, was also a rough draft of his book, a multimedia tie-in already half complete.

Seth ran a triple backup protocol, pushed himself away from his station, and let out a long yowl of triumph, punching the air.

"Initiating Mental Health Protocol II," Tasha immediately responded.

Seth laughed. "Good grief, no, no, cancel that. I'm fine. Just very happy."

"Are you certain? My logs indicate a period of intense, silent activity, followed by severe agitation. These are outside your normal operating parameters."

"Very sure! I've honestly never felt better." Well, that wasn't quite true. He ached all over, still awash in the aftermath of muscle overuse. Seth stood up to do lots of slow arm and leg stretches.

Far from feeling drained, though, he was ecstatic. His brain still felt almost warm, as if he had just solved a challenging puzzle in a short amount of time.

He allowed himself a minute or two of self-congratulation. After getting his implant, he'd spent several fruitless days trying to record his brain during a good writing session. The self-monitoring had rendered him unable to write anything at all. Then he'd had a brilliant idea.

After Tasha had assured him it would be safe to do so, he had recorded his brain while he was in the shower. Twenty minutes of his favourite meditative activity, where some of his best ideas had come to him. And it had worked.

No more stress over the best use of time. No more teeth grinding when his family demanded time from the unmarried guy with no kids. From now on, it didn't matter how long he had available to write, it would now all be optimal.

Seth headed for the kitchen, his stomach growling again. He was tempted to have Tasha print him up two meals at once. Maybe he'd have one now, go out for a jog, and come back for another meal.

He picked something that looked inviting from the fabber menu and stretched again. His mind was racing with the possibilities. At this rate, he'd finish his book by month's end. He wondered whether he shouldn't set aside the draft, start a new book, and then return to the original with fresh eyes. He thought about being able to take weekends off writing, to stay fresh, without feeling guilty about not trying to squeeze out another five hundred hard-won words.

The aroma of his meal, a fresh pizza just like his mother used to print, floated up to his nose. Seth thought about using the same recording for learning book marketing. He hoped to hand the legwork over to Tasha once he'd figured it all out, but there was a lot to learn first. Dreams of selling more than a handful of books danced through his head.

He reached into the cupboard for some utensils, and just then, his wristband beeped. It was his employer. He happily wondered if they had a new Farming assignment for him. Seth felt like he had the energy for anything.

Then he read the message and dropped his fork with a clatter. He had just lost his job.

HAROON

It was late June. It had been baking hot for three days, with so much humidity that stepping outside felt like walking into a thermal therapy bath. And suddenly the room felt just as stifling.

"What do you mean, you're pregnant?" Haroon blinked stupidly.

Saba just looked at him from her seat at his kitchen table, her eyes suddenly wet with unshed tears.

"How can you possibly be pregnant? You told me you had the birth control implant!"

"I did," she said, looking away. "I turned it off."

"You... shut it off?" Haroon couldn't believe what he was hearing. "What? Why?"

They had been dating for nearly three months now. Pretty, smart, and even more well-read than he was, she'd had ambitions to become a nurse, and maybe a doctor after that.

"I thought you loved me," Saba said, crying now.

"I... do. At least, I think I do," Haroon said. "I don't know. I'm eighteen! You're eighteen! We don't know anything about anything! Nobody gets pregnant that early. What on earth were you thinking?"

She sniffled for a long time and then said. "I won't go back. I can't. I thought you'd be happy."

Haroon rubbed his face in agitation. He thought about his RCMP application and how it would look if he withdrew it now. The plans for going to Japan together he and Yoshi had been tossing around. His father and his background, and the fact he didn't know his mother and wasn't sure what kind of person he was, much less what type of person he'd produce in a child. They didn't have the money for genetic enhancement. Was screening covered in their healthcare coverage? His fists clenched and unclenched.

He wanted to scream and rage that he'd only just gotten free.

Haroon sat across from her, trembling. He couldn't quite make himself take her hand in his, but after a while, he said, "Help me understand."

Saba nodded and wiped her nose. "It's Pakistan. It's always Pakistan for him. I can't understand why he bothered leaving. He should have stayed there." She pressed her fists into her eyes. "He thinks war is coming again. He wants to go back to fight, and we would have to go back with him. My mother won't say no to him." She paused. "I-I think she's afraid to."

Haroon thought of her father, a man even grimmer and more brutal than his own. They'd met only once so far. A veteran of the last India-Pakistan conflict, he'd fled the country when he thought he'd be swept up in the UN peacekeeping missions and forced to pay reparations. Saba didn't talk about it much, but there were hints he might be guilty of war crimes, too.

"So, you thought if you were pregnant, he'd want to leave you here," Haroon said.

"Yes… especially if he thinks it could be a boy. And maybe my mother and sisters, too. If I said I didn't know how to have a baby or look after it… which I don't."

"How… how far along are you? Are you sure?"

"Eleven weeks," she whispered miserably.

"Eleven weeks! That was…" He tried counting backward, his thoughts in a riot. "Our first time! Saba!" He couldn't stop himself from blurting out his next thought. "You talk about love? Do you actually love me? Or was I just a convenient donor?"

She gasped and pulled away from him, crossing her arms over her chest, hunching as though he had punched her in the stomach. "How could you—?" She stopped and gulped, tears pouring freely down her cheeks. "Of course I do. You were the nicest boy I'd ever met! You're cute, you're funny. Clever. I-I thought we could make it work."

"You could have asked first!"

"What would you have said?" Saba demanded.

"I don't know!" Haroon shouted. "We'd only just met!"

"That's not true!" she yelled back. "We'd been out on dates before that!"

He stood up and kicked the chair. Saba flinched in a way that made him burn with shame. Haroon turned away from her, his chest heaving.

They were silent for a long time. Then she said, "You don't know what it's like, growing up in a war. Wondering if the sound you hear overhead is a commercial solaplane or a recon drone. Whether when you open the door it will be your sister back from a secret trip to school or a killbot that's followed you home. Or even if the house you left that morning will still be standing. We moved seven times before the peacekeepers finally got to our region. Do you know how much of this is pure luxury?" She waved her hand, indicating his apartment. "Education is free here. There's an income to provide the basics while you're having your kids and strong laws to protect jobs. Universal health care. And we're not constantly going to war! I only just now started to sleep through the night! I didn't mean to rush us... but when father announced we'd be going back..." Her breathing was now ragged, convulsive. "I didn't plan it, not really. I swear."

Haroon wasn't listening. He was thinking of his father and wondering now how he had ended up with a child he hadn't wanted. He didn't know what a father was supposed to be like. He remembered the regular beatings his father had given him, and how he had cowered and cried as he was pummelled.

Only now, when he looked up into his father's face in his memory, it wasn't Subhan's face he saw, but his own.

"I need out," he said, and stormed off, leaving Saba weeping in his kitchen as he slammed the door behind him.

PART III

KEL

The rain was nearly horizontal; the wind gusted and howled, rattling the door. The summer thunder made the windows vibrate.

Kel ignored the noise. She was sitting by herself at a corner table in an old-time bar on Bloor St. called, appropriately enough, Tradition. She was sipping saffron vodka, straight up. It wasn't something she liked overmuch, but it had been her grandmother's favourite drink, and the perfume of it brought back nice memories.

The bar featured real wood, probably the original stuff from when the place was built, and genuine draught taps that had to be pumped by hand. It smelled of stale beer and oil soap. A real antique, the place was, and it appealed to Kel today for reasons she couldn't explain.

It was her third or fourth drink of the afternoon. She had a pair of EduTain's AR glasses on, and to her eyes, it looked like a news-program anchor was sitting at the table right across from her.

"And finally," he said in a pleasing baritone, "Toronto police have announced they are officially looking into those very strange claims involving Casino Max we brought to you yesterday. In case you missed it, we interviewed several people who came to the TorNews flagship program Public Outcry to complain they had all attended Casino Max last week, had won big and then realised long after they had returned home from the casino that they had not received the payout they were sure they had won. For their part, Casino Max officials claim that, in fact, there weren't any big winners last week at all, much less several."

The announcer leaned forward, looking earnestly into Kel's eyes. "Now, far from being a run-of-the-mill gambling scam, TorNews' investigation of the matter revealed some odd details. All the people involved claimed to have won roughly the same amount of money at the same type of machine, and none of the claimants appear to be connected to each other. Many of them were incensed to hear about similar complaints, suggesting a rumour mill had prompted several copycats to cash in. Police say they have no reason to suspect impropriety at the casino and are treating the matter as being a drug-related incident. However, their spokesman, Constable Bryer White, admitted he had no clues yet on what drug could cause the same hallucination in so many people. We'll update you as more details become available."

Kel groaned, pulling off the glasses and letting them clunk onto the table top. She could guess what had happened. Someone at the casino had tried handing out complementary implants and drinks. The addictive dopamine surge associated with what it must feel like to have a big win would be sure to keep customers coming back to lose money, wouldn't it? Except they hadn't realised the experience would feel so real those same customers would feel they had been ripped off, somehow.

The bartender, an older man with a slight paunch who looked like he would be just as comfortable behind a bar in the 1800s as he was today, sidled up. "You know, lady," he said, as he filled up her glass. "Ordinarily, I'd be concerned about you coming in here outta the blue and drinkin' this much while lookin' so glum. Never a good sign when people are doin' that, eh? But I gotta say, I'm just happy to have the custom."

Kel looked around. She hadn't noticed, but the place was empty, and now she thought about it, it had been ever since she came in "Where is everyone? Does this place just not fill up until the evening? Or is it the weather's been too awful?"

The bartender looked disgusted. "Oh yeah, oh yeah, the weather's a thing. But it's this new gadget on the path. Supposedly, it gives you the same buzz as the booze, only without the liver damage and the hangover. And you only have to buy it once. You can stick it in your head, play the same buzz repeatedly. Business has dropped like a stone." He used his bar towel to give her table a wipe it didn't need. "Kids today," he huffed and went back to doing whatever it was bartenders do when they're not serving drinks.

Kel buried her face in her hands and cursed. She didn't know whether to be appalled or depressed or both. She had designed the replay implant to record flow so people could do more good things, and every day now, she was hearing about people using it to simulate highs, or drunkenness, or for pornographic purposes. And cheating, or whatever it was the casino owners were trying to pull off. Why could no one see the true potential of the device?

Her leg ached. She reached down to rub it. The leg, and her post-assault recovery were her official reasons for asking for a short sabbatical, but really, she had left work because she couldn't settle down. When she wasn't obsessing over news about the device's uses, she kept trying to think of a method to kill the spread of the implant before someone got hurt with it. The irony of it all was killing her: she couldn't concentrate long enough to record a good session to help her focus for longer periods on solutions.

Kel thought of a virus for the implant software, but she didn't want to risk something going haywire while it was influencing people's brains. She also didn't know where people were getting copies of software. The devices she'd found so far already had the software installed. Whoever had stolen the original must have put up a publicly accessible copy of it that was pulled as soon as the device was replicated. And the implant was hardware as well as software; a physical object that could be printed by anyone with access to a fabber…

A fabber virus! Kel thumped the table when the thought came to her. That was it, yes. Something to infect the fabber network so the device wouldn't print. Or it would print a dud so enthusiasm for the thing would wane. A passing fad. A weird footnote, like so many others, in Toronto's history.

Her face flushed hot with shame. Memories of how she'd foolishly stolen the first copy of the device she'd found on the flow – in broad daylight, in plain view of the shopkeeper – kept looping in her mind. On top of everything else, she was worried he had video footage of the incident would be made public. How would she ever live it down?

Kel decided she'd have one more drink and then go home. She felt she owed the bartender that much.

SETH

Visiting the Royal York was always like stepping back in time.

Seth took a moment to admire the grandeur of the hotel with its marble floors, dark wood-panelled ceilings, and massive chandeliers. He was fond of the green analogue clock that graced the centre of a spiral staircase in the lobby. It felt like gentlemen in tails and ladies in flapper dresses would appear at any time, sipping swanky cocktails and discussing that Picasso character.

He took the lift to the main mezzanine and made his way to the Algonquin room for the opening reception of an annual writers' conference that was always good to be seen at. The room was already half full, with writers from all over the city making full use of the hors-d'oeuvres and drinks tables.

"Seth!" someone called out. He turned to see a small group of people chatting and sipping. He recognised a few from a short course he'd attended late last year. Seth grabbed something of his own to sip and joined them, surprised to discover he was eager to socialise.

The woman at the centre of the group, who he recollected as being a popular fantasy author, was wearing an enormous hat with a swoopy brim and black gloves. He thought her name might be Monica and he remembered she had a particularly dry sense of humour.

"Seth, darling, how are you?" she said. "I haven't seen you out at a thing in ages. Where have you been hiding? You remember Amachi, Bahram, and Kyra, I think. And this gentleman is Marty."

As they all gave each other little bows, Seth realised the three he had met before were each wearing something very similar to what he had seen them in before: Amachi was wearing all black, Bahram had a thing for lots of gold jewellery, and Kyra had a distinctive facial tattoo. They all had a trademark style. He wondered if it had anything to do with their sales, as they all outsold him regularly. He felt annoyed with himself for not having noticed this sooner. Maybe he was seeing it now because he was still riding the high from his breakout work session.

"Marty was just telling us how he was working on his first book," Monica said with a twinkle in her eyes suggested she found it rather amusing. "Go on, Marty, tell him what it's about."

Marty beamed, then took a big gulp of his drink. "It's about this small town in Ontario; I follow it over several generations, and it's all about how the town can't escape its past and keeps repeating it. And all the characters are zombies. And! Here's the twist: they all have the same name."

"Wow," Seth said, burying his face in his drink for a minute to cover the expression he wanted to make. "That sounds…" Seth searched for a suitably neutral word, "intriguing! Are zombies a thing again? I'm afraid I haven't been keeping up."

Marty's face fell. "What do you mean again? Has my book been done before?"

Seth resisted the urge to tell him all plots had been done before. "Uh no, not that I'm aware of. Sound like it will be very challenging to write. Can't wait to see it published!" he finished, sounding insincere even to his own ears.

Monica rescued him. "So Seth, where have you been?"

"Oh, you know, working on this and that. I have my fourth book nearly done now, I think," he said, keenly aware of the jealous look Marty was giving him. "And, I was doing some, ah, character development for Xperience before they folded."

"Wow!" Amachi said. "Is this the same one you were thinking about at the workshop? That's tremendous. I'm still in the planning stages of mine."

"Nearly done? I remember workshopping your opening scene with you. Weren't you thinking of this one being a pretty hefty epic?" Bahram asked."Change of plans?"

"It will be pretty big," Seth nodded. "It will come in around two hundred thousand words, I think."

Amachi dramatically faked a spit take. "Good lord, Seth. I am trying to do the math on that in my head, but whatever it works out to, it's a ridiculously high number of words per day. What's your secret?"

"A new implant, actually," Seth said, feeling proud of himself. For once, he felt like he had figured out something other people hadn't yet.

"Oh, not the thing I keep hearing about?" Kyra said. "The weird brain playback gadget?"

Seth nodded. "Yeah, that one. I managed to catch one of my best creative sessions with it, and now I just fire it up when I'm ready to sit down."

"Wow," said Marty. "What a great idea. I would pay money for that. I mean it."

"Me, too," said Monica. "Seth, are you selling copies of your session? I find it so hard to put away the distractions these days. Too many good VR things to get lost in."

Seth grinned, delighted. "That depends; how much do you think it's worth?" They laughingly shouted out numbers at him and he could hardly believe his luck. They were all much, much higher than what he'd been thinking. He pointed at Marty. "Sold, to the highest bidder. And to the rest of you at that rate, too, if you're serious." Everyone started poking their wristbands to initiate transfers.

When they were done, they toasted Seth. "Are you staying for the whole conference?" Kyra asked him.

Seth considered his options. He'd only planned to stay for the reception, to say hi to some familiar faces; only a few sessions scheduled for tomorrow had looked interesting. But he had a hunch that word of his productivity tool would get around. It might not be a bad idea to stay accessible. In fact, it might be lucrative. And since he was unemployed and now in debt for all his new equipment…

RAY

When he walked in the neighbourhood now, there was more than deference in people's eyes. There was unmitigated fear. He heard whispers that he had the ability to inflict terrible nightmares on people at will. More than once, he had glimpsed someone crossing himself or making the sign to ward off the evil eye as he passed. It boggled his mind that people would do that around *him*. Around Ray, that little kid that had taken years to work out he didn't need to be afraid of his mother any more. And in this day and age, too. How could people be so superstitious in the midst of all this smart technology? He didn't know it was what Dominic was saying, or whether it was the sight of the 'guests' staggering out of the plaza that was doing it; he suspected it was both.

Stranger still, it seemed like even the guys on Dominic's crew were giving him space. Not showing fear, exactly, but finding ways to leave the room whenever Ray appeared. Big, rough guys who must have seen it all by now.

Not that it mattered. He didn't have the time or the energy to talk to any of them anyway.

Dominic was on a tear. On slow days, Ray would run through his routine at least ten times, each replay just as horrifying and as gruesome to watch in action as the first one. Through snatches whispered into the victims' ears, or the bits and pieces he heard during the rare times he was let out of the room to eat, he had worked out that Dominic was pushing further into the fabber network. He was selling black market fabber recipes, stuff the average Joe on the flow wasn't supposed to be licensed to print. He was bullying his way into politics, taking payments to intimidate certain candidates into calling off their bid for office. His hackers were slicing into big companies to use their big server CPU cycles for encryption cracking. Ray was sure he'd heard someone say something about trafficking in illegal mods and printed body parts for the overseas markets.

At night, in the dark and under his bed covers with a dim light, he made more notes, piecing together a bigger picture of the crime and corruption in the city. The company he suspected of killing Mick was right in the thick of it, with a history of suspicious industrial accidents involving Analogue workers from J-District, and ties to at least two rival gangs. The question was, why Mick? What had he ever done to them? And which person had ordered the hit?

Lack of sleep was making it hard to think. Dominic was bringing in anyone who had ever resisted him in the slightest, or hinted they might resist him, even members of those other gangs. For all their toughness, they fell to pieces just as quickly as the others Tomasso brought him.

Ray rolled his shoulders in his quiet room, shook his head to stay awake. He was desperate not to be caught sleeping on the job, but it was so deathly silent in here when he was alone and waiting, and there was nothing to do but try to avoid thinking of the contorted, terrified faces he had just seen. Or worse, speculate on how those replays had been made.

The door swung open, and Ray stood slowly, moving with the unhurried, preternaturally calm pace he had developed in his role. Dominic walked in alone.

"Ray, Ray, look at you, my good soldier, all by yourself, waiting for your next assignment. You must be starving. Come, eat with me," he invited.

He followed Dominic up the stairs into the bar. It was pouring rain outside; it might have been the fifth day in a row they'd had rain. Ray wondered how waterproof the downstairs area was. The places he'd stayed in as a child had always leaked: water, cold air, awful smells.

They sat on the barstools. Sylvie must have heard them, because she came out almost immediately. Her eyes went to Ray; instead of the friendly smile he'd seen the last time, she swallowed and glanced nervously away. Dominic snapped his fingers at her, a frown on his face, making his demand without saying a word. She vanished to fetch their meals. He stared after her, tapping a rhythm on the bar top, thinking. Ray tried not to fidget.

Dominic turned to him. "Raymond. You've turned out to be quite the asset, did you know that? Just how valuable? Go on, ask me. Ask me how valuable you are."

"How valuable am I?"

Dominic's hand lashed out, catching Ray a ringing blow to the ear. He fell off his bar stool and smashed his head on another stool. Before he could even grunt, Dominic was there, hauling him upward, dusting off his clothes, and inviting him to sit down. Ray saw stars. The old anger surged in him, and the humiliation. He shook with the effort of suppressing the urge to attack Dom.

Dominic took his own seat. "Not priceless. Not irreplaceable. Remember that. But very, precious. Check your account."

Breathing slowly to calm himself, Ray pulled on the chain he wore around his neck to reveal his fob. He checked his account balance and discovered a much bigger number than he could have imagined.

"Thank you," he said. He remembered the crying girl that had begged for relief and wondered if he would ever spend any of the money without thinking of her.

"A token of my appreciation," Dominic said, slapping him on the back. "Our little venture is turning out to be much more effective than I had hoped. That man you did this morning? Government. I now finally have my line on virgin digital IDs. I'll build government-certified profiles to create all sorts of upstanding fake citizens. I'll make a fortune. These things?" He tapped Ray's port. "These things are the key. Raymond, did you know that monkeys, when they want to be threatening, will stare at you open-mouthed?"

"I didn't know that, no," Ray replied.

"They do. And the more teeth you can see, the more threatening the stare. They lunge. They make their hair stand on end to seem bigger." He pulled an implant from his pocket. "These are my teeth, Raymond. They are helping me to scare all the other monkeys. So much neater and tidier and more civilised, don't you think? Not like the old days when the only option was to gun people down in the flow or leave something nasty on their doorstep to get results."

Ray, despite the buzzing in his ear, risked a question. "Does this mean you don't put out hits then? Just the thingweb stuff?"

Dominic paused in the act of bringing a glass of water to his lips and fixed Ray with a cold, hard stare. "Why, Ray? You're not thinking of leaving me, are you? Because no one leaves me without my permission."

"Of course not," Ray blurted. "That just seemed..." Ray searched for the right word. "That seemed more traditional than I thought you were."

Dominic chuckled and took a drink. "Traditional. That's one way to think of it. I try to avoid it. It's always... ridiculously complicated. Very hard to keep quiet and untraceable. And sloppy. You don't want to be messy, do you Ray?"

Ray felt a chill run up his spine. Disappearing later was going to be more difficult than he'd hoped.

HAROON

Haroon met Yoshi at Kaiten Noodle downtown. The garish colour scheme hurt his eyes, but the food was cheap, and that would be increasingly important in the days to come.

"*Yo, tomodachi*." Yoshi grinned. "Good to see you in person. I was thinking we'd become message-only friendies."

Haroon pulled him into a hug. "Sorry brother," he said. "I just needed some breathing space, you know?"

"Fair enough, fair enough," Yoshi said, and they took a table. The place was getting busy, so they placed their orders right away. "So how are you doing? Have you talked to Saba?"

"I'm going to see her tonight," Haroon sighed. "It will be a horrible conversation."

Yoshi gave him an incredulous look. "You're going to split up with her?"

Haroon bit his lip, trying not to cry. "I—I don't know what to do, Yoshi. I've made a child. I should stay with her. I want to help. But—"

Their drinks rolled up on the conveyor belt. "Well, I was pretty certain you'd stick around, but…" Yoshi took a sip. "This is kind of a big thing. I'm not sure I'd have the guts to stay. Can't play tournament-level Outrider if you've got a baby waking you up at 3:00 a.m., right?"

"Yeah," Haroon sipped his soda, not tasting it. "She's going to expect me to have some sort of master plan to make it all work out; only, I got nothing. This would be a long, hard slog. I'm not even sure how we would do this without her father killing me."

"Do you love her?" asked Yoshi.

Haroon was silent for a beat and then nodded. "Yeah, I think so. I figure that's why I was so angry with her when she told me. Felt like she'd betrayed me. Or she hadn't trusted me enough to ask about getting married. Or talk to me about her father and his plans." He suppressed a belch and took another drink. "I suppose you're gonna tell me love will make everything all right?"

Yoshi laughed. "How many romance holographs do you think I watch? I know real life is just a little bit different." It was Yoshi's turn to be quiet. Then he said, "What's it like, anyway? To be in love?"

Haroon swirled his drink around in the glass. "Different than I expected," he said. "All that stuff about feeling giddy, and sweaty palms, and weird feelings in the bottom of your stomach? Never happened. There's just this… this sense of calm, really. I mean, before we found out she was pregnant, of course. But kind of like all of the tumblers in a mechanical lock have clicked into place and the door opens nice and smooth and you feel like you've always been together and always will be."

"I'd never be able to ask my dad that question," Yoshi said. Haroon nodded. Yoshi's parents seemed pretty solid together; they clearly respected each other, but he had never seen them be affectionate. Haroon had the sense just asking the question would embarrass them.

"This is what killing me about all of this. I do want to be with her. And the baby. But I would be such a disappointment to her. She'd be better off without me."

Yoshi made a sad face and came over to hug him. "Don't say that. You know it's not true."

Their food arrived, and they spent a few minutes getting it organised to eat. "So what are you going to do?" Yoshi said, between mouthfuls. "I mean, I can't believe you've really got nothing going into this."

"Don't I?" Haroon replied, bitterly. "I figure we'd maybe have her mum and her sisters for support, if they don't disown us. Her dad is pretty archaic about women. I don't have anything more than the basic income, and no family help to speak of. The RCMP would probably be the fastest route to a regular paycheque, but their washout rate is crazy high. The hours on the job would be brutal too. I'm not sure I've got the brains for this."

"So you'd be marrying more than one woman, really," Yoshi waggled his eyebrows at him.

"Ack, I hadn't thought of it that way," Haroon gulped down his soda. "But yeah. I suppose. The thing that worries me most, well, apart from her father killing me, is afterwards. I was looking at the RCMP because some day I wanna go back to J and clean it up. Like, really go in there and clear it right out. It seems it's like this black hole that gets attention only when stuff spills out into the main city. I'd need to rank up, and for that, I'll need more than the undergrad degree. So I'd have to be a green cop and a new dad and try to study at night, for years, all while trying to be a halfway decent husband." Haroon rubbed a hand over the stubble on his chin where he was attempting to grow a beard. "Have I ever mentioned what my home life was like?"

Yoshi gave him a sort of lopsided smile that was both admiring and rueful at the same time. "Yeah, I think it's come up a few times. You're a little short on the role model end of things. But listen, about the studying…"

"Yeah?"

"You know how I said in my last message there were some weird things going on at the tournament I was at last weekend? The leading team suddenly passing out and forfeiting? Turns out they were using these." He pressed half dozen implants into Haroon's hand. "I want you to have them."

"Uh, thanks? Pretty sure I'd be great at passing out without help."

"Give me a minute to explain, Mr Sarcasm," Yoshi glared at him. "I did some asking around. Apparently, these things are good for recording your brain when you're in a smooth groove. You know how I live for those times in the game when everything clicks and I hit every combination at the right time and see every move coming? And how I can't do that if I've not slept well or I'm hungry or whatever? These things allow you to replay your best brain. I've been thinking about it and I figure you could use them at night to give you a solid hour's worth of clear-headed study, instead of three hours of struggling and jacking yourself up with too many cups of coffee." He reached over and tapped Haroon's forehead. "Just promise me you won't be like those fools and use them for several hours straight. Apparently, those players I mentioned had been for days with them, practising, and then used them all that day, in a high pressure tournament, too."

Haroon shifted in his seat, frowning. "But these look like they need a brainjack and I don't have one—"

"I know, that's why—"

"And I cannot let you buy me one," Haroon said firmly. "I can't do it anymore, Yoshi. I gotta be able to make my own way."

"Aaaand I knew you'd say that, so that's why this would be a loan. With interest." Yoshi took a bite of his food. "Besides, how else am I going to say I got a cop in my pocket?"

Haroon had to laugh. "Ohhh, so that's how it's going to be, eh? Cold, Yoshi. Real cold."

"So's your burger. You gonna eat that?"

Haroon looked down, not at his burger, but at the strange implants in his hand. He felt the first flutterings of possibility, and smiled.

MAURA

The local financial press had discovered the replay device.

"I'll take one more question," Maura said, looking out over the sea of journalists and vroggers all pointing recording devices at her. Some of them had excessively bright lights. She picked one at random. It was of no consequence, really. She could guess what the question would be no matter who she selected.

"Nagesa from the vrog *Virtual Reality Toronto*," the woman said. "The TSX will open in about fifteen minutes. Given the documented ability of this device to provide actual reality right inside your head and so completely eviscerate your company's core products, what kind of bloodbath are you expecting in the market?"

There it was. Maura clamped down on a surge of irritation. She loved her adopted country, but one thing always irked her: Canadian-based companies never seemed to get a good headline at home. It didn't seem to matter how well the company was doing, or what it might be up to in terms of philanthropy or community service, the pro and amateur press always managed to find a negative to focus on. And there was a perverse sense of glee in the reporting if a company put a foot wrong or got caught out by something.

Maura chose her next words carefully, knowing the stock brokerages would incorporate an analysis of her statement into trading algorithms even before she stepped away from the mic. "Thank you. First, I think you're overstating the capabilities of this flow device. It can reproduce certain sensations, of course, but it is still in the very early—"

"Won't stay that way for long!" someone in the audience shouted.

Maura pretended not to hear. "And second, EduTain has been aware of this technology for some time, and we have been reviewing ways of utilising it. We see it as complementary rather than competitive. As I think I've said to you before, we're always looking at the big picture."

"You didn't answer my question," Nagesa said.

Maura flashed a smile she didn't feel. "I expect there will be some movement today. There usually is when the market learns about something new and doesn't know how to price it in. But as I said, this is not new to us, and this will settle out before our next quarterly earnings filing. We're here for the long haul. Thank you, everyone."

She stepped away from the podium and headed for the lift, ignoring the additional questions being shouted at her back. As the doors closed behind her, she felt very unsettled. She couldn't shake the feeling she'd badly misread her audience. Her usual way of dealing with most situations was to project calm confidence and a steady hand at the tiller. Now she was wondering whether she shouldn't have gone for more excitement and perhaps even admitting to some uncertainty.

She didn't have to wait long to find out. Pauline met her at the lift door, bearing a tablet that she handed over without a word. Maura skimmed the headlines. Eight stories had already been filed. All the authors had cast EduTain as a stodgy company about to be blindsided by something new and blast.

Maura gave back the tablet, and they went back to Maura's office. She woke her computer and pulled up a screen for the exchange. "That's not good," she said after a minute.

"What?" Pauline asked.

"Significant pre-market activity on the stock," Maura said.

"Shorting it?"

"I hope not," Maura said. Pauline came around behind her to watch the market opening.

The verdict was swift and brutal. Within three minutes of the opening of trading, the stock's value had been cut in half. Within five minutes, it was cut in half again. By the time it finally plateaued, ten minutes after opening, it had lost ninety percent of its value, and it was still falling. Maura wondered what she would do if it was delisted.

She turned off her computer.

"Well," Pauline cleared her throat, came out from behind the desk, and sat down, looking stunned. "That was dramatic." She straightened a little, trying to put on a brave face. "However, what was the old phrase? It's only a paper loss. Once everyone is done panicking, I'm sure it will come back up. Probably right after we work out an approach for the device and announce something."

Maura's mouth twisted. "Wait for it."

"What?"

Maura's wristband beeped. She looked at it and blew out the breath she'd been holding. "The bank. It's not even a personal note. It was automatically generated."

Pauline's hand flew to her mouth. "The bank? What for?"

"Insufficient market cap. They've cancelled our acquisition financing. Boom, just like that." Shaken, Maura stood up. "If you don't mind, I need some alone time just now."

Pauline left. Maura walked on unsteady feet to stand in front of her Van Gogh, hoping to find solace in his orchard.

KEL

Kel glanced up from her work. Someone was banging hard on her apartment door.

She checked her wristband, wondering who it could be. The video from her door monitor showed two men in dark suits. They were looking at the camera. One of them made a motion towards his ear. She turned on the audio.

"Dr Rafferty, we know you're home. Open up immediately."

Kel flicked off the wristband and went back to work. She didn't know who they were, and she was not in a mood to discuss anything with demanding strangers just now. She almost had the fabber virus finished. It was fiendishly difficult work: not only would it have to spread undetected, it would have to be specific enough to kill prints of her device, clever enough to spot mods of her device and kill those prints, and yet not cause other issues.

A surge of anxiety gripped her. She'd seen the report about EduTain's stock tanking because word of the device had hit the financial press; fortunately, it had been just before the Canada Day holiday weekend and virtually no one else was paying attention. But it wouldn't be long before the media started digging for their autumn news cycle stories or some thingweb vrogger created a conspiracy theory about it. Kel reckoned she had maybe a week left before it was truly out of control.

There was another noise at the door, and Kel froze in shock when she realised it was from someone forcing entry. She recovered quickly, transferred a copy of her virus to her wristband, secured her computer, and looked for something she could use as a weapon. Kel cursed the lack of effort she'd put into decorating. She grabbed a brass candlestick from a shelf, an item she'd received from her grandmother's estate. Kel tucked it behind her back. It wasn't huge, but if she could keep it hidden and then be quick, she might injure one or both long enough to get out of the apartment and into public view.

She turned to face the doorway of her living room just as the two men walked in calmly. One of them had his hand conspicuously in his pocket, where there was something large and lumpy.

"Dr Rafferty, are you usually this rude to guests?" said the man who had spoken earlier.

"Guests don't show up unannounced or break in," Kel replied. She wished she had thought to thumb the emergency beacon on her wristband. "Who are you?"

"We'd like to talk to you about this," the man said, showing her a copy of a replay device. His partner remained silent, still holding whatever it was he had in his pocket. It looked like it was pointed at her.

Kel tried to keep her expression neutral. "What is it?"

"Please don't waste our time," said the man. "We know

this is your work, Dr Rafferty. Every fabber unit has a unique ID code associated with it. Whatever it prints has that code embedded in it, and whatever machine scans and prints something has its ID inserted in the print too. And so on and so on." He pretended to examine the implant. "This copy is a seventh, meaning it's a copy of a copy seven times over. The trail is easy to follow if you know how to look. It leads straight back to you. Yours was the first."

"What is it?" Kel said again. "And why are you here?"

"The problem with this is…" he said, walking forward to halve the distance between them; Kel gripped the candlestick tighter and tensed to strike "…it only records twenty minutes' worth of experience, which means you have to play it over again to get into the same state. Unfortunately, it doesn't seem to be possible to get it to replay again very quickly, which means there's an annoying sensation of resetting. Breaks the concentration. We have assumed, based on your publicly available work, that you're also responsible for the invention and design. We don't quite understand how this thing works, but we like it a lot. So we want you to give us an original copy and we want you to fix the design so it doesn't do that. A nice, seamless loop would be ideal. Or better yet, much longer replay times."

In spite of the situation, Kel's overactive mind leapt to several possible solutions. She shook her head, as much to say no, as to clear it. She reminded herself she had deliberately kept the recording time down to a minimum because she didn't know what the side effects of prolonged use would be. "I'm not sure what you're going on about, but

I do know you have no right to be in my home making strange demands."

"Dr Rafferty, we're a government agency and we see tremendous potential in your device for enhancing our military capabilities. The application for improving sniper concentration and accuracy alone means it is your patriotic duty to put this device to work for your country."

"Not a chance!" Kel was horrified. "That thing is not meant to help people kill. And I don't care what agency you claim to be, but my government doesn't go around breaking into people's homes and demanding they hand over intellectual property and provide free labour." Kel wished she felt as certain as she sounded. Surely you only saw this stuff in spy thriller VRs? Now she was regretting not having paid more attention to politics and the news. "Who are you with?"

The man glanced at his partner, who was looking less sure of himself than he had been before. "We're Department of Defence," he said.

Kel's confidence rose. Perhaps she could outbluff them. "And not in uniforms? I hate to break this to you two clowns, but we have a *Ministry* of Defence in this country, not a department. Now I can't tell whether this is some type of con, or what, but you've wasted enough of my time. Get out before I ping the police."

The man smirked. "Not going to happen. My friend here has a damper field on. No signals in or out of anything within twenty metres of him."

Kel tried to gauge if she had room to get around him and

out the door. "Then there will be a repair team here soon, to investigate a dark spot in the thingweb."

"We'll have you out of here long before then," said the man. "You will come with us, and we will have your cooperation."

"Enough," said the other man. He pulled a gun from his pocket, quick as lightning, and fired. There was a muffled bang, and the top of Kel's left leg felt as though it had been struck by a baseball bat. She looked down, disbelieving, as a burning, electric sensation seared through her thigh. Her trousers suddenly felt hot and wet. The leg wobbled and then collapsed under her. She crashed to the floor, losing her grip on the candlestick. It rolled away behind her.

"You shot me!" Kel gasped.

"Goddammit," the first man said, glaring at his partner. "What happened to using a Taser?"

"She probably has a brainjack," the shooter replied. "She's no good to us fried."

Kel lunged at the candlestick. The first man moved fast, planting a sharp kick just above her wound. She yelled and grabbed her leg, curling into a ball to ward off another kick. The man cursed again and fumbled in a pocket. In a haze of pain, Kel took comfort in the thought things obviously were not going as planned for them.

They produced something like a pillowcase, which they jammed over her head. She felt them grab her roughly, one arm each. They hauled her upward and pushed her forward. She pretended to cooperate for a few hobbling steps, and then deliberately fell, tripping him and sending one crashing into his partner. She heard one man thump into something — her wall? — and hoped it was headfirst.

She sat up and wrenched the sack off her face in time to see the shooter's fist before it smashed into her nose. The force of the blow sent her back down to the floor, hard, and she blacked out.

RAY

Tomasso was waiting outside Ray's apartment in a private, unmarked pod. Ray got in, feeling nervous. Something about Tomasso's invitation that morning, the way he pulled him aside, quietly, murmuring in his ear, had filled him with dread. Was he about to have another surprise surgery? Had he screwed up so badly he was about to be offed? He was so close to figuring out what Mick had been doing at the drone company.

Tomasso only nodded at him as the pod pulled away. Never a man of many words, he stayed silent, staring out into the deepening darkness.

Ray rubbed the sweat off his forehead. Even though it was almost 9:00 p.m., it was still oppressively hot and humid outside. He recalled a cartoon VR he'd seen one morning, peering into someone's living room, about how the dog days of summer had to do with some star rising in July, rather than anything to do with panting dogs. He tried looking up through the pod window at the sky as if doing so would help him remember the name of the star, but he couldn't see much through the wash of the flowlights. Ray noticed his knee was bouncing up and down and forced himself to sit still.

The pod took them on a fast drive across the middle of Toronto and out to the east. Past North York, Ray wasn't sure where they were or where they were headed. The pod's dashboard remained dark, probably at Tomasso's insistence.

After about half an hour, they pulled off a main artery and into a tree-lined suburb. Ray gazed with envy at the beautiful homes and well-kept front yards zipping past. He wondered what sort of person he would be now if he'd grown up in one of these houses, instead of the hovels and shelters he remembered.

The pod slowed just as they were about to go around a curve, and pulled into a small layby near a thickly wooded area. A sign declared the space to be Cudia Park, and Ray's heart beat faster. It didn't look like it was a well-lit place that would be filled with potential witnesses at this time of night.

Tomasso hopped out and motioned for Ray to follow. Ray did so, reluctantly, allowing the gap between him and the gangster to grow a little. If he had a gun, it wouldn't help much, but if he had a knife, Ray might stand a chance if he wasn't within easy striking distance.

Tomasso followed the walking trail into the trees, pulling a flashlight from a trouser pocket; save for the small white beam, it was pitch black underneath the canopy. They walked along the path for a while, Tomasso glancing back from time to time to make sure Ray was still following. Then they left the trail, crunching through the underbrush. Their footsteps seemed unbearably loud in the darkness.

They had walked for several more minutes when Ray noticed Tomasso had donned AR glasses. He also realised there was moonlight filtering through the trees ahead and he could feel a slight breeze. After another hundred paces, Tomasso brought them to a halt, and Ray's breath caught in his throat. Now he knew where they were: atop a bluff overlooking the lake, and the view was both nerve-wracking and spectacular. It was a sheer drop of several dozen metres to the beach below.

Tomasso indicated the water with his chin. "Is nice, yes?"

Ray grunted his agreement, still very tense.

Tomasso cast him a sidelong look. "You got balls, I give you that," he said. "I would not have followed me out here. Not without knowing."

Ray took that as licence to ask the obvious question. "So what are we doing?"

The man shrugged. "You are useful to Dom," he said. "He protects his assets. You do nothing stupid, you gonna be around for a while. I got cash — real cash, not Dom's crypto — buried under here." He pointed to a large tree behind them. "I got a sister, she's got a kid. If I get killed, you take it to them, yes?"

The knots in Ray's shoulder muscles loosened. "Oh. Okay, yeah," he said. "How am I going to find this spot again?"

"I give you the GPS and my sister's address," Tomasso replied. "I bring you here tonight so you can see for real, know what it looks like. I check it lots. I add to it. I count it."

Ray doubted he could tell one tree from another, especially in the dark, but he nodded anyway. Tomasso seemed pleased. He reached into a pocket and pulled out a large flask, unscrewed the lid, and passed it to Ray. The smell of cognac punched the air. Ray took a swig and handed it back. Tomasso took a long drink and wiped his mouth on his sleeve. They looked out over the lake again in silence, passing the bottle between.

"So why now?" Ray said after a while, deciding to test things a little. Obviously, Tomasso had decided he trusted him, and maybe he could deepen the connection, and learn more. "Dominic's been making lots of moves lately. Something I should be worried about, something big?"

"Eh, maybe. The Dom is pushing hard — some people don't like it. Most likely, I think I got one strike left."

"Oh yeah?" Ray wasn't sure what he meant.

"Yeah," Tomasso took several more gulps of the liquor. "I mess up twice now. Next one, that'll be all." He drew his finger across his throat.

"You sure?" Ray accepted another small pull from the bottle and returned it. "I see you with Dominic all the time. Everyone knows you're his second. You'd have to do something terrible to make him want to lose you."

Tomasso made a dismissive gesture. "The first time, it was no so bad," he said, his accent thickening as he drank more. "This last time, it was no so good. The Dom is not a patient man." He took a few more gulps, grimacing as it burned its way down. He laughed. "No so good at all."

"What did you do?"

"We kill the wrong guys," Tomasso said. "The wrong guys, can you believe it! Two Joes downtown. Drone assassination, very big mess."

Ray went very still.

Tomasso drained his bottle, and patted his pockets, feeling for another. "I mean, pffft, a couple of dead bodies, who cares? No big deal. And I no so stupid as to have the drone trace back to us." He found his other flask and pulled it out, shaking it to see how full it was. "But the Dom, he was mad we didn't get the guy he wanted."

"This was the thing in the winter?" Ray could barely choke the words out.

"Yeah, yeah, this is the other problem," Tomasso shook a finger as though scolding. "See, you heard about it. He said I should learn what discreet means. Hits should not be headlines, he said. Bah!" He fumbled the lid off the bottle and drank. "So, one strike left for me. Then I get to be your next replay recording." He shuddered and gripped his flask so tightly the sides dented inward.

Ray swayed, and suddenly the noises of the wood behind them and the water below them were too loud and the light of the moon was far, far too bright.

Tomasso rolled up his sleeve. The word discreet had been carved into his forearm. The scars were still red.

Ray saw Mick's dead body, his face a nightmare of melting, morphing features freeze-framing into glimpses of all the people he'd ever worked over in The Room.

The wrong guys, can you believe it!

He saw himself wake up in the hospital and watched as he exploded when he realised what he had lost.

A couple of dead bodies, who cares?

"Not two dead bodies, Tomasso. Just one."

"Huh?"

Ray swung his right fist in a fast, wide arc to the left, smashing it into Tomasso's face. Tomasso's flask went flying, and he stumbled backward, his arms flailing. Ray ducked low and charged, his shoulder digging deep into Tomasso's stomach, slamming him into a tree trunk. The impact bent Tomasso double, knocking the wind out of him, and Ray crouched and then lunged upward, lifting Tomasso bodily off the ground and slinging him over his shoulder like a wet sack.

Who cares? No big deal.

He spun around, took three long strides to the edge of the bluff, skidded to a stop, and heaved Tomasso into the void.

Tomasso bounced once off the cliff face, a shockingly heavy thud against solid earth that reverberated under Ray's legs. Then he fell, end over end like a rag doll, until finally, he was a crumpled, dark heap on the beach.

Retching and shaking so hard he could barely walk, Ray grabbed the flask he had been drinking from and staggered out of the woods. It took him all night to walk back to his apartment.

As the sun rose, he ate a steak Dominic had left him for his breakfast.

SETH

Seth accepted the four sticky-kid hugs without too much grimacing before taking a cleaner hug from his sister Sandy.

"Thanks again, Seth," she said. "That was great. The kids always enjoy time with you, but today was extra special. I think you may have sparked a career interest in Lucca."

They were queued up at the pod beacon station just outside the Toronto Zoo, waiting for their turn at the head of the line where they'd be paired up with a ride home. After the hug, Lucca had resumed his intense study of a book about bears. His younger siblings were engaged in a loud game of rock, paper, scissors.

Seth tried to look nonchalant, but inside, he was very pleased. It had been a long time since he'd been able to play the extravagant uncle role, taking them out somewhere fun, instead of making do with toys at his apartment. And although the kids had gotten increasingly fractious towards the day's end — it was hot and they were all getting tired — it had been fun. He'd been looking for an excuse to visit the Centre's new conservation and citizen science training exhibit since they had built the facilities on the old parking lot area ten years ago.

"Ah good, I'm glad everyone enjoyed it," he said. "Now

I've got more time work-wise, we must do this more often."

If Sandy knew the real reason was money, rather than time, she didn't say. They reached the front of the line, so she gave him another quick squeeze and herded his nephews and nieces into a pod. He climbed into the next one, waving goodbye.

"Destination?" the pod asked.

"Harvie Flow at St. Clair," Seth replied.

"Estimated arrival time, forty-two minutes." The pod accelerated away from the station. "Entertainment options?"

He was distracted by the sight of a swarm of spider buildbots printing what looked like residential units. "Hey, I didn't think they were allowing new housing developments out this way."

"If you are referring to the building project to your right, it is a temporary facility designed to accommodate a large academic conference in December. The finished constructs will be relocated to a development area in Vaughn in the spring."

"Huh," Seth craned his neck to see them again. They looked like they would be nice when they were done. He wondered if he would be able to afford one in a few years' time. A vision of how he might decorate it popped into his head and made him smile.

"Amusement options?"

"Headlines. Maybe arts and entertainment."

"Breakout VR star Meike Bergholtz made a surprise appearance at the Royal Alexandra Theatre to the delight of fans. She announced she would take on a role in the upcoming stage production of *Uncle Vanya* and there would be a VR tie-in allowing ticket holders to role play in the garden of the Serebryakov family estate."

"Fun if you're into Chekhov, I guess. Next?"

"Organisers have said next year's red carpet Argy Awards will be held in Vancouver in June. Competition for the top award in augmented-reality productions is expected to be fierce, as several strong contenders have already been revealed ahead of the usual new releases season."

Seth wondered if there was also an award show for virtual reality productions; he'd never paid attention to such things before now. There had to be a category he could enter that supercut video he'd been thinking of, showing all the times he'd died in his short-lived career as a game farmer. "Next."

"Author Marty Foley, whose debut novel ZOMG shot to the top of the bestseller list in fiction this week, says he has already written a sequel. He promises that it is 'even more twisted' than the first book."

"Wait, what?"

"I'm sorry, I do not understand your query. Shall I proceed with the next headline?"

"No," he said sharply, an uneasy sensation overtaking him. "Tasha," he said, addressing his DPA instead of the pod computer, "have I met Marty Foley?"

"Yes," Tasha replied. "At the conference last month. He was one of several people you sold a copy —"

"No, no, NO!"

"I'm sorry —" both computers began.

"Shut up," Seth said. He tapped the pod dashboard

screen to pull up a thingweb story about Foley. The picture showed him looking more tired and drawn than Seth remembered. It said Foley credited an intense regimen of meditation and exercise, combined with divine inspiration for the story and the speed with which he was able to complete the sequel. Critics were calling the book a 'lyrical masterpiece,' that was 'open to infinite interpretations' and 'brimming with life.'

"It's a book about zombies." Seth muttered. "*Zombies!* Who writes about those things anymore?"

His chest felt tight, and he could barely breathe. It felt so unfair. He was certain Marty had exploited his replay recording to finish his first novel and write the second in record time. How he'd gotten the first one published so fast he didn't know. Right now, he didn't have the heart to figure it out.

Seth pressed clenched fists hard into his legs. Marty had used the one thing Seth had ever managed to sell easily to beat him to the prize. With his first book. It was a gut punch. A mixture of jealousy and rage over his own stupidity made him feel lightheaded. Why had he sold copies of his implant recordings? It had been an advantage and he'd practically given it away. He had a vision of Dario, a disapproving frown on his face, coming to fetch Lucca away from him. In his head, his mother's voice asked him when he would get a real job.

His chest constricted further; he could feel his heart thumping. "Stop the pod," Seth said.

"We are on the Don Valley Flow," the pod nav said. "I

do not recommend disembarking."

"Stop the pod," Seth grated. "I need air."

The pod gracefully swerved left through four lanes of traffic and came to a halt at the side of the flow. Seth got out, shielding his eyes against the low evening sun. The other pods zipped by, creating a constant, gusty breeze. He walked in a daze, looking out over Toronto, thinking of all the people in the city, and probably across the country, who now knew Marty's name. Seth stopped and leaned against the barricade, his forehead pressing into a railing still warm from the heat of the day. Seth felt like he was seven again, seated at the dining room table, desperately thirsty, asking patiently for a glass of water, and never getting any, while all about him his siblings yelled and were shouted at and cuffed, but got what they wanted.

A loud bang made him jump. He spun around in time to see a pod, with a blown tyre scraping against the ground, come skidding towards him.

KEL

Kel woke to a radio report of a fatal accident on the Don Valley Flow, feeling like she was surfacing from some long, horrific dream. She cracked her eyelids open a tiny bit and saw a man a metre away from her, mouth agape, snoring. Nothing around him looked familiar. She closed her eyes again, thinking furiously, trying to understand.

She was moving, she realised. Or, more accurately, she was in a vehicle that was moving. Kel risked another quick peek. She seemed to be lying in the back of a cargo pod. It suddenly felt to her she had been here for a while. She focused on listening to the radio, and after several tense moments, she was relieved to hear it was a CBC station. In spite of the movement, she was at probably still in Canada. Okay, good. Then it dawned on her that if the news had reported on something in the Don Valley, she was likely in Toronto. That was even better.

The radio programme changed from the news to one of the station's regular features, and the announcer said the date was July 17. That felt wrong.

Then everything came flooding back and she had to suppress a gasp: she was a captive, and had been so for more than a week.

Kel tried to piece together what had happened. She remembered being shot and being dragged out of her apartment. Then there was… a gap in her memory, because then she had come to, and she had been… here. Yes, she'd been here all this time. The two men had seen her awake and again ordered her to hand over plans for the Implant and provide improvements. Kel had unwisely told them to go to hell, and the first man had stomped hard on her gunshot wound. Another blackout.

She had brief flashes of lucidity after that. Once, she woke up sweating, feeling as if she was on fire. Kel had demanded treatment, reminding them she was no good to them dead of an infection. The man who had shot her had dumped a beer on her leg, causing her to scream and swear. The other had at least fabbed some horrible-tasting antibiotics. The men had argued… a lot. Yes, nearly every time she'd been conscious, she'd heard them arguing between themselves. There were also excruciating memories of being dragged in and out of the pod to use a bathroom, but where? How had she not been seen?

Kel opened her eyes again to study her captor. He was slumped in the corner of the pod now, oblivious. While before he had looked, at least superficially, like a mysterious and very dangerous government agent, now he looked like an average guy. He was dressed in ordinary clothes, and his shirt had what appeared to be a drink stain down the front of it.

Kel tried moving and almost cried out. Several days of lying on a hard floor and a gunshot wound had left her feeling stiff and weak. Keeping one eye on the man, she ever so slowly propped herself up on one elbow, grinding her teeth to stifle the sounds she needed to make from the pain. She was still in the outfit she had been wearing when she was abducted; one trouser leg had been cut off, and her thigh was wrapped tightly with a dressing and sticky gauze tape that looked like it might have come out of a first-aid kit. Well, that was something at least. She didn't know what a bullet wound was supposed to feel like, but she figured it must have gone straight through, or else she'd be in even worse shape.

There was one overhead light in the cargo bay but no windows, so she couldn't tell whether it was night or day. The floor was littered with takeout containers, far more than a week's worth. Some of them looked like they had started to biodegrade. Kel was suddenly conscious of all the smells in the small space: her own, unwashed body, the smell of him, and the pungent odour of spoilt food. She tried breathing through her mouth.

Kel spotted a toolbox and a mobile fabber, which made her think this might be a roving repair station rather than a cargo pod. That would make sense if they were travelling in circles around the city, as it would be a good way to hide. She remembered the gunman's damper field and wondered why no network had yet noticed there was a moving dark spot on the grid.

She glanced up at the walls of the cabin and stiffened in shock. A portion of one wall was covered in printouts of photos of her. There was one of her at the bar. Another of her going into the hab building late at night. Another was of her stepping out of a pod outside of her apartment. The hairs on her neck rose and prickled.

The other printouts were even more chilling. One was a lengthy screed on how the lunar base was supposedly a hoax: people approved to travel there were secretly selected by the government for elimination and then forced out an airlock. Another explained how digital tattoos were supposed to be a United Nations mind-control technology that would be activated soon, turning everyone who had one into an obedient soldier of the Illuminati. Half of the other wall was a complicated chart dedicated to tracking of the movements of the Prime Minister. By the time she was done looking at everything, Kel was terrified.

She was in the hands of two people who were certifiably insane.

Fighting rising panic, she returned her gaze to the man and flinched when she saw him upright, staring at her with red-rimmed eyes, his hands clenched into balls in his lap.

RAY

Ray was numb.

It was now three days after the night at the park. Not knowing what else to do, but sure he couldn't run away — not yet — he had come into work every morning since, only to be told to go home by midmorning. At home, he lay on his bed and stared at the ceiling.

This morning, as he came down the stairs at the plaza, he was aware of a buzzing, angry tension. He knew instantly that meant they had found Tomasso. Ray sighed. He should be petrified, he realised, but could only muster a painful, throbbing ache in his chest.

He walked into the lion's den.

Dominic, who had a commanding view of the entrance, saw him and motioned for him to come over. His eyes were darker than Ray had ever seen them. "Ray, Raymond, *il mio mago*. It is good you are here."

"Is it?" Ray said.

"Yes, because you are not out there," replied Dominic, pointing up at the outside world. "It is about to get very dangerous. You will be safer here."

This was not the answer Ray was expecting. "I don't understand."

"Of course you don't, poor Raymond. Nobody tells you anything."

Ray suddenly noticed there was a newly printed gun on Dominic's desk and everyone around him was heavily armed.

"Tomasso is dead. The *Sǐwáng* have told us to take it as a warning."

"What?" Ray knew from his research — his now laughable research — they were a triad gang trying to reorganise in Toronto; but he also knew they hadn't hit Tomasso.

"No one," Dominic said quietly, "takes out my second without… a penalty." The steel in Dominic's voice caused Ray to twitch. Dominic saw it. "Oh, don't worry, you don't have a long line-up of *mago* work ahead of you. We went out last night and took out several of them."

"But…" Ray felt like the ground was shifting beneath him. Was their blood on his hands, too? "But the *Sǐwáng* is…" He struggled to find a way to question the claim without giving himself away. "They're small. Would they have dared?"

Dominic's eyes narrowed, and Ray tensed. "Interesting. You're more observant than I realised. Yes, I am aware they likely didn't do it. But the fact they would try to take credit for it deserved a response. No, it was almost certainly Posse Wild. We hit them too. Even harder."

Ray's mouth opened and closed. He slowly shoved his hands into his pockets to steady himself.

"You're worried, I can see that." Dominic smiled, a shocking hint of kindness in his eyes. "I agree, I wish we'd been able to keep it civilised. Doing hits is so last century. But you know, two steps forward, one step back." The faintly maniacal look returned. "Although I have to admit last night was a rush. Old times are sometimes good ones, and anger … anger, my dear Raymond, has such a purity to it."

The room swam in front of Ray's eyes. He remembered the adrenaline surge when he had charged Tomasso and before, when he had raged at the hospital attendants and the many times before that, as a young kid bouncing from shelter to shelter, blind fury at the world fuelling his strength. There *was* a kind of purity in those moments. Had he been stupid to leave the district? He looked at Dominic. Was this what he was meant to be?

There was a thunderous crash upstairs, followed by a scream — Sylvie! — and several fast popping sounds.

Dominic grabbed his gun. "Here?" he roared. "In my house?" He vaulted over the desk, heading for the stairs.

But before he could get there, an ear-splitting boom rocked the whole building. The ceiling exploded downward, debris flying everywhere. Ray dove for cover behind a basement support column. He heard feet hitting the ground heavily… one… two… three… four… and then the room erupted with automatic gunfire.

Shards of cinder block went flying as the bullets strafed the column. Parts of the wall disintegrated as dozens of rounds penetrated the drywall, tearing it to pieces. Someone came running out of the bar room, gun ready, only to be mowed down, sliding, collapsing in a heap near Ray's feet.

A round smashed into the light above his head, and it blew up with a bang and a shower of sparks. Ray felt needles in his scalp as he hit the floor beside the dead man, sprawling flat, and facedown. His shirt was warm and wet.

The gunfire kept going, punctuated only by soft grunts and the terrifying sound of bodies thumping onto the hard concrete. Then, just as suddenly as it began, it stopped, and all he could make out was the last few bullet casings *tink-tinking* to the floor in a gentle shower. Ray stayed very, very still.

"Which one was Dominic?" said a man's voice, in a heavy East-European accent.

"There, I think. In the fancy white shirt."

"Milosh, stand cover. Vojin, and you and you, check we've got everyone." There was a movement. "Hey, Dominic, eat this." There were three more gunshots in quick succession. "Now you no do your brain sorcery on me or nobody else." More movement, and the sound of a fist striking dead meat several times. "*Kopile.*"

Ray felt, rather than saw a man brush past him. The body beside him got a sharp kick, and he braced, holding his breath in the hopes he wouldn't give himself away if was kicked.

But the man moved on, and Ray slowly exhaled, quivering.

There was muffled conversation in the distance, the sound of guns being dumped, and then footsteps receding up the stairs.

Ray waited as long as he dared. If he moved too soon, he

could be shot by anyone still watching the scene. But the warm wetness had spread up on to his back. What did it feel like to bleed out?

Mustering his courage, he rolled sideways and sat up in one scrambling motion, crouching behind what was left of the column. No one shouted. No one fired at him. He risked a peek from his hiding spot.

There were bodies everywhere. The pipes in the ceiling, broken in the explosion, were gushing water straight down onto Dominic's desk. There was already a large, red-tinged pool rolling steadily towards him.

He stood up and pulled off his shirt, patting himself all over. Ray couldn't feel anything — no holes, no broken ribs. He glanced down at the body he had hidden beside and saw his own outline in the puddle of blood seeping out from the other man. He put his hand to his head; it came away sticky and glittering with glass fragments. Ray nodded to himself as though it all made sense.

The stench of toxic smoke reached his nose, reminding him of the garbage can fires of his childhood. Emergency services would be here any minute.

All of a sudden, he felt bone tired. For a long moment, he surveyed the carnage, imagining what it would be like when the police came charging through the door upstairs and found him. He saw himself kneeling, hands on head, giving up.

Giving in.

No more struggle. Accepting his fate as some anonymous thug, destined for jail since the day he was born.

He climbed gingerly upstairs on the ravaged staircase.

It was hot up here. Through the doorway, the booth where he had once slept was on fire, and flames were licking the bar top, making the varnish bubble and stink.

Ray threw his bloody shirt at the fire and watched it sizzle and curl and then vanish.

His fingers found the fob on the chain around his neck. He pulled hard until the chain snapped, and looked at it in his hand. It held a fortune in cryptocurrency.

Ray tossed it in the fire and saw it burn.

Yes, he could give in.

Or he could try, just one more time.

He found the back door and slipped out.

KEL

"I'm getting tired of waiting for you to smarten up," the man said. He reached into the toolbox and pulled out a heavy wrench. "I don't care what he says anymore. We've got work to do, and you're just in the way now. And you've cost me a fortune in food."

"Why didn't you explain it to me?" Kel said, swallowing hard.

The man, who had started towards her, stopped, looking confused. "What?"

"All of this," she said, gesturing at the walls. "Why didn't you tell me? I had no idea all this was going on."

The man looked at the walls, frowned, and looked back at her.

"Look, I'm sorry for being rude and stubborn before," Kel continued. "But I'm a scientist, you see. I required evidence. You should have just explained this at the beginning."

The man's eyes widened in surprise as he processed this.

Kel gave him some time before deciding he needed another nudge. "What else don't I know about?"

Still eyeing her suspiciously, he reached into a box she hadn't spotted until now and pulled out what looked like more printouts. He shoved them at her.

Trying to keep her breathing in check, she made a show of inspecting them one by one. There was a piece on how induction-charging plates in the flows were designed to produce waves to control people's minds. Another lengthy printout discussed the chemical content of the water supply.

"Unbelievable," Kel said, and meant it. "How long do you suppose this stuff has been going on? Right under our noses!"

Now the man nodded emphatically. "You see it, don't you? Now you know why we need your tool. We've got to get rid of the heads of the hydra!"

"Okay, let's get started." She extended a hand. "What do you want me to do first? Here, help me up."

He reached out and she took it, making as though she was getting to her feet. At the last second, just as the vehicle curved into a turn, she heaved on his arm, pulling him off balance and heaving him forward to smash headfirst into the wall behind him.

Groaning with the effort, she rolled over and grabbed desperately at the wrench he had dropped. As he rolled onto his back, clutching his head, she scrabbled on to her hands and knees. She gripped the wrench with both hands and raised it overhead, brought it down hard onto his groin. He screamed, sat bolt upright, and then curled over into a tight ball of misery.

Still keeping a firm grip on the wrench, she hauled herself up, feeling dizzy and faint. She put a hand on the mobile fabber unit to steady herself and surveyed the back door of the pod. She didn't know how to stop the vehicle but she also had no idea if she could keep him incapacitated long enough to figure it out. Her last move had been as much luck as quick thinking.

She glanced down at the hard surface under her hand. A fabber unit. Kel still had her wristband on. She might not get another chance to upload her virus.

Working quickly, she unlocked her band and popped it open to reveal the universal hardware port that still came as standard with every band in case over-the-air connections failed. Kel plugged it into the fabber and instructed it to download a design schematic for immediate printing. The design resembled her implant prototype. Embedded in the fake were a set of calculations that would unlock and extract a virus payload. The virus would make a copy of the fake and send it across the fabber network to as many nodes as it had direct links to. From there, it would try to print itself, starting the cycle all over again and, more importantly, it would copy itself over any of the real prototype schematics resident in memory. From now on, Kel hoped, anyone who tried to print a new copy of her device would just relaunch the virus.

And, she prayed, not mess up the entire network.

The man groaned and writhed. Muttering under her breath, she made impatient hurry-up gestures at the fabber. When she was sure the virus had launched, she clipped her wristband back on and limped painfully to the back of the pod, leaning on the wall for support. She grabbed the interior door handle, pushed the release button, and shoved.

The door swung open, flapping as the pod zoomed along. It was dark outside, and there weren't any pods behind them. Kel guessed they were on a service flow route and it was late at night. She gulped. It looked like they were going very fast.

Her legs were shaking with exhaustion. Her captor would not stay down all night, and she couldn't stay upright for long either. It was now or never.

She jumped.

MAURA

"You did what?" Pauline looked appalled.

"I said I arranged financing for the takeover using my personal assets as collateral," Maura replied.

"But," Pauline sat heavily in the chair by Maura's desk, "that means if the company fails, you're done for."

"Yes," Maura said.

"Why would you risk that?"

Maura resisted the urge to tell Pauline she'd asked herself that twenty times since this morning. "It's a calculated risk," she said, as much to reassure herself as well as Pauline. "It's a tactic to shore up our third quarter."

"We're just about to file our second-quarter earnings. They look as solid as ever because they were all in the bag before news of this implant hit. Yes, I see…" Pauline nodded. "If we announce a takeover bid at the same time as we put up solid results, it could bring the stock back up."

"It will demonstrate confidence in our future far better than my botched press conference did."

Pauline hesitated and then said, "Don't be so hard on yourself."

Maura grimaced. "There's no sugar coating that one. I blew it. They needed something more than the bland pap I offered them. If I hadn't done that, I wouldn't be in the position I'm in now."

"If that researcher hadn't created the implant, you wouldn't have had the press conference in the first place," Pauline countered. "You must hate technology, sometimes. It can flip things upside down at a moment's notice. Your whole life's work is on the line here."

"Hate it? No, not at all. Look around you. You can't tell me living and working like this," she pointed at her office and the city outside the windows, "isn't infinitely better than trying to eke out a living by farming out on the prairies like the first settlers did. If the mosquitoes and the blackflies didn't drive them insane, the long winters stuck in a one room soddy did. No, technology is not the issue. It's still the answer to many problems."

"No, the real difficulty is some humans in particular and humanity in general." Maura sighed. "The people who invent things or discover things usually have good intentions, and they think what they've produced will make things better than they were. And it's true, to a point. But they hardly ever take human nature into account and try to predict how something might be abused or simply misused and then put in measures *ahead* of time to mitigate that. Most of the recent history of science and technology has been about launching something and then rapidly backpedalling and applying bandage solutions."

Maura got up to look out the window, gazing across the city to the university's main campus. "I do not know what our doctor friend over there had in mind when she designed those implants, but she obviously didn't take the time to secure her prototypes, and she was naïve about the level of privacy she had when on her computer."

"Not to mention the device's ability to play back just about any state, it seems," Pauline said.

"Is that what our lab is saying?" Maura asked, turning to face Pauline.

"Yes, as far as we can tell there's no limit to the experiences it can record and replay," Pauline said. She rose, stretched a little and joined Maura at the window. "So, inventors and scientists should be made to study more what, history? Philosophy?"

Maura laughed again. "What, you mean if I were to become prime minister or something and could decree this? Those wouldn't hurt, and neither would psychology. But actually, the problem goes further than that."

"How so?"

"Well, very few of us are good at reasoning things through to their logical conclusion. Or even just playing with the idea 'what if? When I said humanity that's what I meant. We're terrible at long-term thinking. We really need to teach the skill."

Pauline appeared to be very amused. "Well! That explains why we maintain a few product lines that aren't super profitable. That's what those are, aren't they?"

"Yes. Those speculative pieces. They're thought experiments. Each book, each VR experience, each holographic is a mini-laboratory. Helps you think through possible futures. My favourite genre." Maura shrugged. "So those lines are my pet projects. You know, I've never told anyone that before."

"What, not even your best friend?"

Maura's face clouded, and she turned back to the city view. "I've never made any time for friends," she said, "My parents seemed to have all kinds of them when I was growing up. I think I told you they liked the good life? Lots of parties, always having people over. Nobody stuck around after my folks were killed. Just poof, gone. My cousin took me in but only because I was the heir to the company he got to play with until I reached the age of majority. Didn't seem much point to friends if they vanished on you. And besides, I had to deal with the other parts of my parents' estate and get my education, and then I plunged straight into running the company and resettling it here, where I knew no one."

Pauline placed a gentle hand on Maura's arm. It was warm through the cloth of Maura's suit jacket. "I'm sorry," Pauline said. "I can't imagine how that must have felt."

Maura's face brightened. "My goodness, that's given me a great idea!"

"Oh?"

"Yes! And it'll guarantee we stay afloat. We need to get in touch with Dr Rafferty again."

PART IV

KEL

Kel rolled over and groaned. Her head was on fire, and every muscle in her body felt like it had been stretched like a rubber band and snapped violently back into place. And coming to in strange places was getting very old.

Then she smelled smoke.

Panicking, she tried to sit up, but the movement made the world spin, and then all she could do was throw up. She stayed like that for several minutes, alternately puking and gasping, until she felt that she'd heaved up everything she had ever eaten in her life.

Moving much more slowly, she sat lay back down and opened her eyes.

She was outdoors… somewhere, under a shelter of sorts. It was pouring rain, the water coming down so hard and in such volume, that she could only see a few metres ahead. Kel was sure there'd be flooding in parts of the city again. The smoke was rising from a small fire. Beyond the flames, a woman regarded her warily.

"You'll want water," the woman said. She reached behind her and produced a small bottle. "Just now, all I'd do is rinse."

Kel accepted the water gratefully. She took her time

sitting up, and to wash the foul taste out of her mouth. She looked around. "Where am I?"

"Old Dowling Avenue Bridge," the woman replied, staring into the fire. "I come out to the lake for a swim now and then, and if I'm too tired to walk all the way back, I stop here for the night."

Kel twisted her body to see. This was indeed an old bridge, overgrown with summer weeds. She looked more closely at the woman. Her hair was dark brown, shot through with grey and pulled back into a messy bun. Her face was tanned and shockingly wrinkled. Her clothes were, to Kel's astonishment, handmade.

"You're an Analogue!" Kel blurted out.

"The name's Byela," the woman said, frowning at her. "Never have been too fond of that term."

"I'm Kel. How did I get here?"

"I found you by the side of the flow. I normally don't get involved in other people's lives, 'specially when they look all beat to hell, but, well…" Byela flicked a glance at Kel, looking embarrassed, "you were wearing one of them new augs," she said, indicating the back of her head. "And I got nosey. So I played it. Whatever you've been into lately was some kind of terrifying. And then I brought you under here."

Kel wondered what part of the last several days had been recorded. She hadn't turned on the record function, but maybe one of the two men had. Then she looked again at the woman in her rough clothes. "Wait? You played it? But how?"

Byela gave her an odd look. "Through my jack, of course." She lifted her hair, exposing the implant underneath.

"But… how can you afford that?"

"Just what do you think an Analogue is, exactly?"

Kel paused, unsure of herself. "People who can't afford…?"

Byela looked disgusted. "Oh my, just where have you been all this time? So all Analogues are just poor bastards without access? All of them desperate keen to get on the marvellous thingweb if only they could?"

"But…"

Byela sighed. "Look, here's the way it is with me: I love the outdoors. Always have. And I want as little as possible separating me from it. So about ten years ago, I gave up my career and disconnected as completely as I could."

"You *chose* to live like this?"

"Yes, goddammit, and another reason is 'cause I got tired of people like you spoutin' off about how exponentially better things are and inventin' stuff to make things even easier for anyone on your side to live but not paying attention to the real problems we still got to solve." Byela grabbed a stick and poked a log in the fire, pushing it higher and making the flames jump. "You know what finally put me over the edge? When someone put out software to help track all of my various fabber prints so I could post my statistics on my connection networks. Not to help me reduce resource use, but so I could brag about my numbers. And this was before it looked like we were getting better news about the climate. So I said, that's it, I'm out. I'm not participating anymore."

Kel wondered at the woman's logic, as she was still using resources one way or the other to stay alive, but thought better of saying so. "Why do other people disconnect?"

Byela threw the stick on the fire. "Other folks have political reasons, some have religious reasons. Most probably don't even have a clear reason for staying away from augs, or fabbers, or whatever. They do because they want to. And yeah, there's lots of them who want to connect but can't." She pulled up her sleeve to reveal her digital tattoo. "But here's another thing. I can get back in any time I want. Do you have any idea how hard it is to get one of these things if you're not already connected?"

Kel frowned. "It can't be that hard. There are government agencies and—"

"And how do you know about those if you haven't grown up with them? How do you navigate a world no one's ever taught you about? Those are skills you need to be taught."

The question had never occurred to Kel. An ID was just something that had always been there. Head down, hard at work, eyes on her goals and her career. She'd always had a place — a node — in society, and connections on the thingweb, and had never questioned why or how that was. She pinched the bridge of her nose, feeling headachy and tired. So far, this whole year had been a crash course in human nature, one she was failing badly.

Byela was on a roll now. "And these latest implant things. What good are they supposed to be? People usin' them to get their rocks off? Pffft. Like humanity needed more ways to do that."

It was more than Kel could take. "That's not what they were for!"

"Oh yeah?" Byela's chin jutted out stubbornly, not happy to be interrupted in a good rant. "How would you know?"

Kel thought of all the things she'd seen on the news, and all the other uses she'd worried about in the wee hours of the morning once she'd seen how creative people were being in abusing it. She didn't want to feel responsible for all of that. This was not what she'd envisioned when she'd daydreamed about becoming a famous neuroscientist. A muscle in her jaw clenched. "Because I invented it."

Byela was momentarily speechless. "So what's it for then?" she finally asked.

"I made it," Kel said, "to help people get better work done—"

"Oh geez, another 'productivity aid.'"

"No!" Kel shouted. Byela tensed and glanced at the stick she had tossed into the fire. Kel took a calming breath. "Not like what you're thinking, anyway. I made it so we could reproduce that state of 'flow' on demand. That feeling of being totally synced up so we could put that kind of intense creative energy to use on tough problems. Only it all went horribly wrong. Someone got a copy before I finished it and it got out, and people have been using it for all kinds of crazy things I never would have dreamed of. People have been hurt because of it. It's an awful thing."

"Oh," Byela said, mollified.

They both stared into the fire, listening to it crackle and pop. Then Byela said, "It's not awful, you know."

Kel just looked at her.

Byela shrugged and looked back into the flames. "Truth is, I'm also out here 'cause I don't like people very much. Misanthrope, I think the word is. I avoid people as much as I can. And like I said, I found you by the flow. Saw you had one of these, and well, I don't know what came over me. I'd heard about these things and I took out your chit and played it. I'm sorry. But then I couldn't unfeel it. You'd been through hell, all that fear, and pain, and I really felt like I'd been there with you. I could tell there was big rain coming, and I brought you here to… I'm not sure what. Keep you dry and safe until you could use a beacon for emergency services, I guess."

"Empathy," Kel said, wonderingly. "You felt what it was like to be me, and it moved you."

"That's what I'm saying," Byela nodded, her cheeks colouring. "And if you can get a cranky hermit like me to make nice, well, that's somethin', isn't it?"

Kel had to admit it was.

MEIKE

In her dressing room, Meike unlocked a large custom case she'd had fabbed a few weeks ago. Inside, there were dozens of implants, each with a neat label in a tiny font. She ran her finger across the second row, stopping to pull out the one she was looking for. Meike reached around, pulled out the existing implant, and popped in the new one. She was just filing the original when there as a knock at her door.

Lorenzo came in, an angry expression on his face. "You did it again, didn't you?"

"What?" Meike said, not looking at him. She sat in front of her mirror to do her makeup.

"Don't give me *what*," he said. "They told me you did your own stunts again yesterday."

Meike pouted into the mirror. "It's more fun that way. And it's what got me into every entertainment roundup as a headliner. No one does stunts anymore, much less their own stunts. It's all computer generated."

"I know," Lorenzo fumed. "And it was fine as a one-time thing to make your name. But I don't want you injured."

"I didn't know you cared that much," Meike said, applying foundation to her face.

"You're worth a lot of money to me. More alive than maimed or dead."

"Such a sentimental man."

"Pah! Just knock it off already. Your amazing acting skills are all you need now."

"It may not last forever. That reminds me," Meike said, sticking out her hand and making a gimme motion. "What else you got for me today?"

Lorenzo glared at her, but Meike just made the gesture again. He sighed, stuck his hand in a pocket, and grabbed a handful of implants. He separated one out and held it up, grinning. "This one is jealous anger."

"Where'd you get that one?"

"Off Julie, right after I told her I was screwing you, too." He pointed to a tiny cut near one eye. "She threw it at me."

Meike rolled her eyes at him. "And you think doing my own stunts is dangerous? You're lucky she didn't push you down the stairs." She took the implants and looked them over. "What else is here?"

"One is a bad LSD trip. The other is off a guy from his first time base-jumping. Not sure how useful it's gonna be. But you said you wanted to cover as many situations as I could think of."

Meike held the base-jumping session implant up and contemplated it. "What's the date today?"

"I think it's the twenty-eighth?"

"We're doing a chase scene. This'll be perfect," she said, tucking it in a pocket. She filed the rest in her special case and locked it. "Now out. I've got lines to memorise."

He grabbed her arm and pulled her close. "Hey now, you know what I just told Julie. Don't make a liar out of me."

She pushed him away. "You're already a liar. I'm not the only other one. And if I don't get to work, you don't get paid. Which do you value more? Paid or laid?"

"Oooh, cruel choice," he laughed. "So I won't make it. I'll go hook up with someone new while you work."

He was still laughing as he ducked the hairbrush she tossed at him while he was leaving.

SETH

"Mama, you can stop fussing, I'll be fine," Seth said, feigning exasperation. "I'm being well looked after here."

He was in a private room in Toronto General, with the late summer sun streaming in the window. He was propped up just enough to see his mother, sitting in a chair by his bed. As far as he could tell, she had been there every day since he'd first been brought in.

"Nobody can look after you as well as your own mother," she said firmly. "And frankly, Seth, I can't wait to get you out. They say you'll heal much faster in familiar surroundings, surrounded by your people. We'll put you in the spare bedroom for a few weeks." She poked him gently in the ribs. "You need feeding up, too."

He suppressed a wince as his ribs were still sore. "You'd have me move back home? I don't want to be any bother—"

"Seth Carrado Bacchi, we all thought we'd lost you," his mother said sharply, her voice cracking. Seth looked at her, astonished. Her eyes were wet with unshed tears. He'd never seen his mother show anything other than zen-like calm and control, even when they'd been rowdy, crazy children. "Did you hear that, Seth? Did you know the reporter who was first on the scene got it wrong? They told us you had been the one to die?"

He took her hand. "Yes, a nurse told me. I'm sorry, Mama."

"That part wasn't your fault. Although what possessed you to go for a walk on the parkway I will never know," she said, some of her control returning. "So the least you can do for scaring us all to pieces is let us love you to better health. Your nephews have been beside themselves, and Dario will be here when his flight gets in this afternoon."

"Dario is flying all the way back from Africa?" Seth blinked back his own tears. "Just to see me?"

"Of course he is, Seth," she said, looking at him as though he had a head injury on top of everything else. "Why wouldn't he be? He's your brother, after all, and he loves you."

"It's just…" he started to say and then stopped. All this attention was astonishing.

His mother waited for him to finish, and when he didn't, she sighed. "My poor Seth, always the quiet one. So good with words on paper, but not as loud as you need to be in your big, shouty family." She patted his knee. "I need a drink. A hard one. Can I smuggle something in for you? What would you like?"

He was taken aback by being asked a preference. "You know, I'd kill for a soda. Maybe a ginger ale?"

She winked and got up. "I'll pop over to the store across the flow to get some. I'll grab some rye to go with mine."

Seth watched her go, feeling tired. Was it possible he'd read them all wrong all these years? Was it just a matter of shouting louder in his family? He had a feeling it wasn't quite so simple, but it was all too much to contemplate right now. What he did know for certain was that wandering out of the pod had been a damn fool thing to do. It had been a sulk that had nearly cost him his life. Never again.

He had just closed his eyes to wait for his mother to return when he heard someone come in. He opened them, and to his astonishment, discovered Marty Foley standing there, twitching with impatience and barely controlled anger.

"Uh, hello?" Seth said.

"Was it a setup all along, or did you decide at the last minute to take me out?" Marty demanded.

"What?"

Marty stamped forward and threw a paper on Seth's sheet-covered stomach. "Whatever it was, it worked. Don't tell me you haven't seen this yet."

Frowning, Seth picked it up. It was a printout of the most recent bestseller list. His mouth fell open when he saw not one, but two of his titles in the top ten. Then he shook his head, not daring to believe it. "Is this some sort of joke? If so, it's in really bad taste, given the circumstances."

"So you're denying you set this up then?"

"Set what up?" Seth asked.

"This," Marty said, waving his hand wildly at Seth's hospital bed. "This over-the-top publicity stunt of yours. I knew established writers didn't like it when young guns came up the ranks, but I never thought you'd do something like this!"

Seth could only stare at him. "Marty, I was clipped by a

pod on the Parkway. My hip and leg had to be rebuilt. My arm is broken in several spots from where I was slammed into the guardrail. I have several broken ribs and stitches in places I didn't even know I had places. The only reason I'm still alive is because it was a glancing blow. And someone else died! You can't possibly think I arranged this?"

Marty seemed not to have heard. "And the first one on the scene, conveniently, is a reporter," he said, "who not only thought to ID you but also decided to look you up." He paused and pursed his lips. "I gotta say, I wish I'd thought of it. 'Critically acclaimed author killed in bizarre pod failure' is a brilliant headline. Crazy viral. How much did you pay her for it?"

Seth hoped his mother was buying more than just one serving of rye. What kind of person thought anyone could be capable of such a thing? His eyes narrowed as he regarded the righteously indignant man. He decided to be direct. "Marty, just how did you do it? Get published so fast and on the list?"

"Wouldn't you like to know?" Marty said. Then his vanity appeared to get the better of him. "I struck a sweet deal with a small press outside the city. I guaranteed them a bestseller if they'd publish me. Then I hired a firm to buy up copies. They had an AI program that knew what fab stations to hit to get ranked. I got it cheap." He thrust his hands in his pockets. "Had I thought to do what you did, though, I'd have been able to extend my run on the list."

Just then, a nurse came by and shooed Marty out, telling him it was family-only visiting hours.

Seth relaxed back down into his pillow. Nothing was as it had seemed. His family loved him. He'd been a chump about how fame and success worked all this time. He'd made the bestseller list in spite everything. And he had dozens of story ideas. Seth was glad he'd left his implant at home that day, and it hadn't been destroyed in the accident. He had a great deal of work to do if he wanted to capitalise on his newfound recognition.

And he was still alive. Seth glanced around the hospital room and realised he hadn't worried about germs the entire time he'd been in here.

Happy tears ran down his face. He was still wiping them away when his mother returned.

MAURA

Maura was in her office again, the centre of her universe. The room seemed oddly confining — she felt she had spent far too much time in it. It was a strange sensation. She had never sensed the lack of anything else in her life before; the business had always been enough.

Hadn't it?

Perhaps it was the recent near-death experience of the company that was unsettling her. Maura tried to shake it off and focused on the two women seated across from her, Pauline to her right and Dr Rafferty to her left. She smiled.

"Dr Rafferty, thank you for coming in to see us, especially under the circumstances. Pauline here tells me it has been a rather trying summer for you."

Kel's mouth twisted into a wry smile. "You could say that. A theft, an assault, an abduction, two police interviews, a couple of trips to the hospital, and a press conference from hell."

Maura regarded her shrewdly. "Out of all this though, is it safe to say the press conference was the worst part?"

Kel looked as though she would protest, but then nodded reluctantly. "The public speaking part was fine, I've done that before. But it wasn't easy answering questions about my prototype. It's been used for so many things. I feel responsible." A muscle along the side of her jaw worked. "I am responsible."

Maura made a demurring noise. "Only to a point, Dr Rafferty. You can hardly be blamed for how other people decided to use something they weren't meant to have yet. I'm confident, left to your own devices, any implant that got released properly would have had safeguards built in."

"I don't know," Kel said honestly. "I'm not sure I would have thought of everything I needed to. People were… very imaginative."

Maura conceded the point. "Maybe, maybe not. People can be very enterprising when it comes to their immediate, short-term interests, and not so concerned about their longer term interests. Sometimes that seems very irrational, and you strike me as an eminently rational person. But in any case, we're all a lot wiser now, I think. I believe you might have a proposition for us?"

Kel sat up a little straighter in her chair, looking somewhat combative. Maura couldn't blame her. Rumour had it she had been besieged by offers — and threats — since the press conference. And Maura had also heard that new copies of the implant weren't functional; they wouldn't print properly and schematics were mysteriously disappearing. Watching Kel now, she suspected she was behind that somehow. Maura had set Pauline on a mission to quietly buy up existing implants before people realised the problem.

"I want to partner with your company to integrate my technology into your virtual reality experiences," Kel said.

"With several caveats. First, I need to test it properly and thoroughly, to see whether regular use is safe. Second, a user would have to review a tutorial and specifically agree to have it as an add-on to their experience. Third, I would need to have built in safety measures, limiting the time they could use an implant, at least until I determine what longer-term operation does."

Maura felt the thrill of new horizons. She struggled to maintain a calm, dispassionate exterior. "Well, that's interesting and most serendipitous. We were thinking along those same lines."

"Yes, well," Kel frowned. "From the beginning? Because that's another caveat: you have to tell me whether it was you who stole it in the first place. Otherwise, no deal."

"That's fair enough. No, it wasn't us."

Kel's brow furrowed deeper. "But you were monitoring my computer, right?"

"Not yours specifically," Maura admitted. "Traffic for the whole institute. We won't be the only ones, by the way, especially now you're famous."

Kel contemplated that. Then: "What sort of experiences would you be considering?"

Pauline leaned forward. "Straight out of the gate," she said, "we want to partner with some of the national charity organisations to boost fundraising efforts. Imagine the uptick in revenue if you could get people to feel what it was like to starve, or what it felt like to hide from enemy bombs?"

"I've become cynical enough in the last several weeks to wonder whether it wouldn't devolve into disaster porn," Kel said. "People trying it for the emotional kick."

Maura nodded. "There will be people who would want to use it like that. We think a baseline fee per use, which would be, in effect, the lowest-level donation, might mitigate the problem. After the experience, they'd be asked to make a bigger contribution."

Kel was thoughtful. "What else would you use it for?"

Maura consulted a list. "Inspiring experiences. Things to motivate people to stretch themselves. What does it feel like to be a music star, singing in front of a large crowd, communing with the audience? What would it feel like to win an Olympic medal? Can we have people enjoy the thrill of an extreme sport without hurting themselves?"

"And on the medical front," Pauline chimed in, "what if a psychologist or a doctor could briefly experience what a patient is feeling to better diagnose and treat them? Or if not the human doctor directly, the med AIs they run? The implant's output could be read and analysed just like any of the other monitoring devices we routinely use these days."

Kel's eyebrows shot upward. "That has a bunch of ethical implications we would have to work through. But, wow, that's interesting."

Maura glanced at Pauline, who nodded imperceptibly. They'd both noted the use of the word 'we'.

"Just the tip of the iceberg," Maura said. "I'm sure there would be more positive applications, particularly if we used your device the way it was intended. In a state of flow."

"Yes, to brainstorm ideas," Kel agreed. "Empathy. It's what we need most these days. Not just for someone you are close to, but a person you've never met."

"Indeed," Maura said. She noticed Kel was leaning back now, her hand stroking her chin. "I think it's clear here we're both headed in the same direction. I can only imagine what we could accomplish, all of us together in a room and all plugged in. We'd definitely like to license the technology properly and have you officially on board as an advisor and consultant, with an appropriate salary. And whatever you decide, we will make a donation to help fund your main line of research, Dr Rafferty. We feel it's the least we can do."

Kel stood up and stretched out to shake Maura's hand. "Thank you. I'd like to think about what you've suggested here, and fully flesh out the ideas and possible consequences."

"Of course, take your time," Maura said. Pauline walked Kel to the door and then returned to her seat.

Maura steeled herself for the next part of her agenda. "And what are we to make of you, Pauline?"

"Me?"

"Yes, you. I can't figure you out. So I am asking directly." Maura sat back and steepled her fingers. "For the longest time, I've had to wonder if you were a plant. Certainly, my troubles seemed to start not long after you came on board. But for the life of me, I haven't been able to trace you back to another company. And your actions when you're with me have all been positive and helped me get *out* of trouble. But there's the matter of your snooping through my office." Pauline flushed red. "So then I thought you were playing a complicated long game. Frankly, it's driving me nuts and I'd like to have done with it. So here we are."

"Oh," said Pauline, still quite pink. "Oh, my. I…" She swallowed before continuing. "I'm sorry, I didn't mean to cause you so much consternation. I suppose it's the case I wasn't being entirely truthful when you first asked me why I wanted to work for you."

"I'm listening," Maura said.

"This is… this will sound weird. But… I saw you on the news a few years ago. I don't remember even what you were on it for. You were talking, and everything you said and the manner in which you acted really clicked with me. I felt like I could relate to you and… well, I just wanted to become your friend. In addition to wanting to learn from you, like I said. The only way I could think to do that, given your position and wealth, was to come work for you. I mean, otherwise, we don't travel in the same circles. I grew up in a working class family."

That wasn't the answer Maura had been expecting. She didn't know what to do with it. "Really? That's it?"

Pauline nodded. "I'm so sorry. I had meant for things to evolve organically, and if it didn't, I was going to go get another job. I—I should just go."

She got up to leave, but Maura raised a hand. "The office? What were you looking for? That's what made me think you were a spy."

Pauline reddened even more. "It was—just a spur of the moment thing. I wanted a really good look at the art you had chosen, to see if I could learn what made you tick, you know? It never occurred to me that you'd be monitoring your own office." Pauline put her face in her hands. "This is so embarrassing. You must think I'm a social climber or some calculating, gold-digging careerist."

Maura remembered a VR star she'd really liked as a young girl, before she'd gotten into the business, and how wonderful it would be if they could have been friends. She was struck with a feeling of admiration for Pauline's nerve. And, she realised, a sense of relief. The thought she might have to fire Pauline had been bothering her more than she'd allowed herself to admit.

Maura tapped the arm of her chair. She wasn't sure what she felt about of any of this. "I see," she said, aloud, finally. "Well. This is all very awkward. And it changes … well, I'm not sure what it might change. But…" Maura sighed. Where were her VR dialogue writers when she needed them? "Look, it won't be long before the patios have to be packed up again. So why don't we call it a day and have a drink?"

KEL

Kel arrived back at her office that morning and found Bao-Yu being handcuffed.

"What the——? What's going on?" she asked Robert, who was standing there watching the arrest, his arms folded and a grim expression on his face.

"Can't say," he replied. "Personnel matter. You know how it is."

Padraig sidled up to her. "Well, *I* can say, as I'm not the boss. It seems we have a saboteur. She's being arrested on mischief charges. Several birds in the hab have been poisoned."

"Nothing's been proven, Padraig," Robert admonished him. "Don't gossip, please."

Bao-You rolled her eyes. "Spare me. You have me on the camera, you said so yourself. You've probably already had a screening of the video with your favourite staff."

"You!" Kel couldn't help exclaiming. Both Robert and the arresting officer glanced at her, frowning.

"Do I need to know something, Kel?" Robert said.

"I'm sorry, with all the … with all that's been going on for me this summer, I didn't have time to bring it up with you." She stopped, unsure what she should say now, with everyone else in the room.

"Go on," Robert said, glaring at Padraig. "He apparently knows it all anyway."

"Well, it's just I had some macaque deaths earlier this year that seemed… unusual."

Padraig leaned over conspiratorially. "I knew it all along! Bird poisonings, monkeys being killed, stuff being stolen. I was sure something was up!"

Kel just gave him a look, and for once, he had the good sense to subside. Kel turned to Bao-Yu. "But why? And did you steal my prototype?"

Bao-Yu gave her a scornful look. "If I'd known about your toy do you think I'd have bothered with your furry brats?"

"But why?"

"In China, there were eight thousand applicants for my job. There were twelve thousand for my supervisor's job. The only way to get anywhere is to have multiple degrees, many years of experience, and several augments. And you glide in here to a top position fresh out of school, get a big research project, and you don't even do so much as take nootropics. I needed your posting on my resume, and what I got was a temporary visiting contract. I thought I would do better here, with less competition."

Kel shook her head. "The birds though? They weren't even mine?"

"Mine," said Robert. "My guess is if she couldn't get you fired, making me look incompetent enough to warrant a transfer elsewhere?"

"I still don't get it. Surely, with all the sensors and

cameras here you knew you'd get caught? How would that help you?"

"It was worth the risk. If you hadn't gotten distracted by that replay device, a couple more dead monkeys and you'd have been fired for negligence or I could have made it look like mistreatment."

"Enough now," the arresting office said, nudging her toward the door. "Dr Rafferty, perhaps you'd like to visit us later for a statement? It sounds like you have something to add. I'd take you with me now, but under the circumstances it would look too much like a perp walk for you and I would guess you don't want the extra publicity."

Kel flashed back to the intense press conference, and shuddered. "No, thank you."

The officer and Bao-Yu exited. Padraig dashed off to spread the news to other staff members, leaving Robert and Kel staring after him.

"I kinda liked her," Kel said, at last. "What will happen to her?"

Robert shrugged. "Not a lot, honestly. If you have proof about your macaques, that will put the charges to mischief over five thousand dollars. She may be made to pay reparations. She'll have her work permits invalidated and be sent back home. After that? Who knows? Bao-Yu seems resourceful, I'm sure she'll land on her feet."

"I guess I had better go to the station," Kel said.

"Yeah. And take the week off. Hit the beach, read some books. Come and see me on Monday." He looked a little sheepish. "This whole thing makes me realise I haven't been doing a great job at mentoring you. I guess like Bao-Yu, I figured someone as smart as you has it all worked out. But I can see now you could have used more support, more career help. I should do better there."

Kel nodded, feeling even more shell-shocked now than when she had come in to see Bao-Yu being arrested.

It had certainly been a year for surprises.

HAROON

Haroon waved his wrist over the pad and then pressed his thumb into it. It beeped to confirm his signature.

The recruiting officer, a man named Max this time, bowed in welcome. Haroon returned the gesture. "And with that, you're on your way to Depot and twenty-six weeks of training," he said, beaming. Saba reached over to kiss him on the cheek, and Yoshi pulled him into a bear hug. Haroon's answering smile was bittersweet. In spite of everything, he was a little sad his father wasn't here to see him sign up. He wondered if he'd ever stop feeling that way.

"They try to take preferences into consideration when setting up your first posting," Constable Max was saying. Haroon was amazed it was assumed he'd pass the course. "And I hope you'll head straight back here. We've just had an anonymous informant come forward with all kinds of details about activity in the District, and I know that's something you wanted to get into. Several notebooks worth. It will take months to verify and investigate properly."

"Thank you, sir, I will," Haroon said, fighting rising anxiety. It was all happening so quickly now. He stuck his hand in his pocket and felt for the implants Yoshi had given him, fingering them as if they were good luck charms. Saba, sensing his tension, squeezed his other hand. He looked at her closely, noticing her quiet strength and determination for the first time. Haroon eased his shoulders back. Maybe they'd be okay.

He accepted a transfer of travel details to his wristband and left the RCMP office. Yoshi went home to practice for another tournament. Haroon watched him go a little wistfully, then put an arm around Saba, and set a beacon for a pod.

From one ordeal to the next: It was time for a lunch date to tell Saba's family of their plans to get married.

RAY

It was now late afternoon. The sun was warm on his face and his upper body. He looked up at the trees nestled between the university buildings and watched the tops of them move in the slight breeze. Nearby, a small flock of starlings was busy pecking at insects in the grass. Somewhere below him, in the shade of the bench he was sitting on, a cricket was thrumming. That sound had always felt like change to him, reminding him that autumn was coming, with its cooler days and soft colours. It made him savour today's gentle heat all the more.

Someone sat on the other end of the bench. That wasn't unusual, as the Walk was busy with students coming and going all the time. He glanced sideways and then did a double take.

It was *her*.

They sat in strangely companionable silence.

"You know," she said, after a while. "I used to see you sit here almost every day, staring at these buildings. Then you disappeared for a while, and now you're back again. I noticed," she said, crossing her legs, "because there's something intense about the way you look at them. It's like you want in."

Ray said nothing.

"You going to apply?"

He shrugged, looking down at his feet. "I don't know. I don't have the other education yet. And I'm not sure they'd have me. I've... done some things."

She laughed softly. "Yeah. Me, too."

They watched people go by. She shifted the package she'd been carrying; Ray saw that it was full of books. He recognised one of the names from the news. She stuck out her hand. "My grandmother would have said everyone deserves a second chance. Maybe that applies to us too. My name is Kel. Maybe I can help."

He looked down at her hand and saw her digital tattoo, achingly aware of the bare skin of his own wrist. He looked up at the treetops, caught the leaves flickering in the bright sun.

Then he turned back to Kel, put out his own hand to take hers, feeling its warmth and the connection.

"I'm Ray," he said. "Ray Tilson."

About the Author

Chandra Clarke wears many hats, sometimes all at once, which makes it hard to get through doorways. A recovering/relapsing entrepreneur, she founded the editing company Scribendi.com, which was acquired in 2017. She will also admit to having been a freelance writer, with publishing credits in places like *Popular Science*, *Canadian Business*, and yes, even *Voice of the Kent Farmer*. Chandra has an MSc in space exploration studies, which she got solely so she could say "As a matter of fact, I am a rocket scientist." She recently finished a PhD, because she's something of an academic masochist. She's a mother to four kids and two dogs, and wife to Terry Johnson, the best British import since the Aston Martin. Chandra thinks her family is pretty awesome, but she might be biased.

If you enjoyed this book, please consider leaving a review where you bought it.

If you would like to see other books and short stories by Chandra, please visit
http://www.chandrakclarke.com/books/

To get announcements about new books, and her regular blog posts, please subscribe at:
http://www.chandrakclarke.com/subscribe/

Printed in Great Britain
by Amazon

39623868R00210